Work the World
with System Fusion

by

Andrew Barron, ZL3DW

This book has been published by Radio Society of Great Britain of 3 Abbey Court, Priory Business Park, Bedford MK44 3WH, United Kingdom
www.rsgb.org.uk

First Edition 2023

Amateur Radio Insights is an imprint of the Radio Society of Great Britain

ISBN: 9781 9139 9532 4

Cover design: Kevin Williams, M6CYB
Typography and design: Andrew Barron ZL3DW
Production: Mark Allgar, M1MPA

Printed in Great Britain by 4Edge Ltd. of Hockley, Essex

Any amendments or updates to this book can be found at:
www.rsgb.org/booksextra

A note from the author
Many of the examples in the book are based on my observations using the System Fusion radios that I own. There may be variations between the equipment that I am using and your equipment, and that may mean that some things mentioned in the book don't work or need a minor adjustment. But most of the information is generic and will apply to all System Fusion radios, hotspots, and services. Where possible I will try to cover any variations. The rules regarding your local repeaters may be different, so it is advisable to talk with members of your local radio club and the local System Fusion community.

Work the world with System Fusion

Other books by Andrew Barron

Work the World with D-Star

Work the World with DMR

Using GPS in Amateur Radio

Testing 123
Measuring amateur radio performance on a budget

Amsats and Hamsats
Amateur radio and other small satellites

Software Defined Radio
for Amateur Radio operators and Shortwave Listeners

An introduction to HF Software Defined Radio
(out of print)

The Radio Today guide to the Icom IC-705

The Radio Today guide to the Icom IC-7300

The Radio Today guide to the Icom IC-7610

The Radio Today guide to the Icom IC-9700

The Radio Today guide to the Yaesu FTDX10

The Radio Today guide to the Yaesu FTDX101

ACKNOWLEDGEMENTS

Huge thanks to the guys and women who give their time for free to run our radio clubs, battle to protect our bands, teach and mentor new hams, make great podcasts and videos, write software, create and maintain the reflectors, talk groups, and rooms, and provide the servers that keep the digital voice modes alive. I know that it is often a thankless task in the face of grizzles and gripes from hams who don't lift a finger to help. Or maybe only one finger. And thanks to you for taking a chance and buying my book.

ACRONYMS

The amateur radio world is chock full of commonly used acronyms and TLAs (three-letter abbreviations :-). They can be very confusing and frustrating for newcomers. I have tried to expand out any unfamiliar acronyms and abbreviations the first time that they are used. I have assumed that anyone buying a YSF radio will be familiar with commonly used radio terms such as repeater, channel, MHz, and kHz. Near the end of the book, I have included a comprehensive glossary, which explains many of the terms used throughout the book. My apologies if I have missed any.

Approach

This book is a practical guide that explains the steps that you need to follow to make your new Yaesu digital radio work on your local repeater or hotspot, and for worldwide contacts. The learning curve is nowhere near as steep as learning about DMR (digital mobile radio), but there are a few new terms to discover, including dashboards, reflectors, rooms, hotspots, Fusion, and Wires-X. Also, acronyms like AMBE+2, DV, GM, PDN, C4FM, and MMDVM.

Operating a digital voice radio is a little more difficult than operating an FM radio, but with a Yaesu radio and a repeater or a hotspot, you will be able to 'work' stations all over the world. I cover accessing the YSF reflectors and the Yaesu Wires-X network, including the 'PDN' terminal mode and the HRI-200 interface unit.

The book includes information about the Yaesu radios that I have purchased. The FT5D, FT-70D, and FTM-400XD. Other Yaesu handheld radios are similar. Hopefully, my procedures should cover most radios. But there might be some minor variations. The FT5D instructions should relate well to the FT3D, FT2D, and FT1D with descending levels of compatibility. I think that the FTM-300 is closer in design to the FT5D than it is to the FT-400XD, but sometimes the FT-400XD instructions may work better. The FT-70D and the FT-400XD seem to be less like earlier radios.

There is a huge amount of excellent information about YSF (Yaesu System Fusion), in the form of online websites, forums, pdf files, and "how to do it" videos. It is not my intention to replace these valuable resources. I hope to complement the insights that they provide and concentrate as much information as possible into a single document.

Many readers will already have bought a Yaesu digital handheld or mobile, but for those that haven't, there is a brief guide to the features that you should look for.

MMDVM (multi-mode digital voice modem) 'hotspots' are very popular accessories. I have included information about their uses and configuration. I cover the new duplex hotspots and the more familiar simplex hotspots, including a section on how to assemble a hotspot from a modem board, a Raspberry Pi, and an SD card. This is followed by step-by-step instructions for configuring the Pi-Star hotspot operating system.

I included a section about using the BlueDV program with a 'DV dongle' which lets you access several digital voice modes without a radio. There is information about the YSF data structure and some observations on the advantages and disadvantages of digital voice technology over FM. This is followed by a brief comparison between YSF and other digital voice modes such as D-Star, DMR, and P25.

The Glossary explains the meaning of the many acronyms and abbreviations used throughout the book and the Index is a great way of getting to a specific topic.

Conventions

The following conventions are used throughout the book.

DV is Digital Voice. **DR** means Digital Repeater. **FM** means analogue FM.

YSF, Yaesu System Fusion, Fusion, or System Fusion all refer to the digital voice standard employed by Yaesu radios. It uses **C4FM** modulation.

FT5D includes the FT5DE (European) and FT5DR (US and International) models.

(R&E) means that the setting or function applies to both the E (European) and R (US and International) models.

Mobile Station. A 'mobile' usually refers to a vehicle-mounted radio, but it can just mean a radio that you use while travelling. In this book, 'mobile' may mean a vehicle-mounted radio or a handheld radio.

Wires-X is a Yaesu-owned and controlled network of servers that provide communal spaces or 'rooms' where Yaesu owners can chat with other Yaesu users worldwide. You can access Wires-X through a genuine Yaesu repeater, an HRI-200 node, or using the PDN or HRI mode on some Yaesu radios.

Wires-X Internet-linked groups are referred to as '**rooms**.' YSF, FCS, and YCS internet-linked groups may be referred to as 'rooms,' or more commonly, '**reflectors**.'

YSF, FCS, and **YCS reflectors** are worldwide networks of public access Yaesu C4FM compatible servers, that are not Yaesu-owned and operated. You can access them via a non-Yaesu repeater, hotspot, or DV dongle.

I have used a **different font** to indicate menu steps in PC software and data entry using the radio buttons.

I use the > symbol to show steps in a menu structure. For example,
Press **Quick > Memory Mode > Group Select**.

Hold means to hold down a button until something happens. The radio might transmit, a beep may be heard, or the display may change. It usually takes about one second. It is a technique that allows a single button to have two functions. A quick press of the button does one thing and holding the button for a second or two does something else.

'**Click**' or '**left click**' means to click the left mouse button, and '**right click**' means to click the right mouse button. 'Click on,' means to hover the mouse over a button or menu option on the PC software, and then click the left mouse button to make the selection.

'**Check**,' '**checkbox**,' or '**checked**,' means 'tick,' 'tick box,' or 'ticked' in UK English.

What is Yaesu System Fusion?

Yaesu System Fusion (YSF) is one of the three most popular digital voice modes alongside DMR and D-Star. Yaesu System Fusion and Wires-X are exclusive to Yaesu. You have to use a Yaesu radio to access Yaesu Wires-X 'rooms.'

Alternatively, you can access thousands of YSF and FCS reflectors using a hotspot, a DV dongle, or a non-Yaesu repeater. Many of these reflectors are in turn linked to DMR talk groups, D-Star reflectors, Wires-X rooms, and other digital voice technologies. The other day on 'HUBNet,' I listened to a DMR user in the USA chatting with a D-Star user in the UK and a P25 user in Australia.

Yaesu digital voice radios use C4FM (continuous 4-state frequency modulation) which is similar to the 4FSK modulation used by DMR and the GMSK modulation used for D-Star. Yaesu digital transceivers and repeaters can swap over to analogue FM if an FM signal is received. I believe that all Yaesu YSF radios are at least dual-band 2m and 70cm radios. Some are multi-band with HF, VHF, and UHF, and others can receive a wide range of frequencies including the aircraft band, FM broadcast band, and VHF marine band. The Yaesu radios are very high quality. They perform very well on digital voice, and FM unlike many cheap DMR radios.

Having three popular digital voice modes leads inevitably to questions like "which is best," and "what should I buy?"

WHAT SHOULD I BUY?

The best advice I can offer is to buy the digital voice format that is most popular in your area. If you have a local Fusion repeater, go for YSF. If most of your friends are using DMR, buy one of those. The same is true for D-Star, P25 or NXDN. There are many cross-connects between Yaesu reflectors, DMR talk groups, and D-Star reflectors, so you can easily communicate with amateur operators who are using one of the other modes.

Yaesu System Fusion is fairly easy to use. Unlike the other modes, no registration is required.[1] Simply set the frequency and offset of your local repeater or hotspot, select the DN digital narrow mode, and press the Wires-X button.

If you are connected to a genuine Yaesu repeater or a PDN or HRI-200 Wires-X node, the search function on the radio will list the available Wires-X rooms. If you are using a hotspot, multi-mode repeater, DV dongle, or non-Yaesu repeater, the search function will list YSF and FCS reflectors. If you want to host a Wires-X room yourself, you need two Yaesu radios and an HRI-200 box, to create an exceedingly expensive access radio and hotspot combination.

Note 1: you do have to register with Yaesu if you want to use the Yaesu Wires-X software to access the Wires-X network directly in PDN or HRI mode or with an HRI-200 interface.

I should point out that although the Yaesu radios are excellent, I found it EXTREMELY difficult to get the ADMS programming software to work. The problem was getting Windows to accept the USB connection to the radio. I had to load the USB drivers multiple times. This is my third book about digital voice radios. I did not have any problems configuring the programming software for DMR and D-Star radios.

The FT-70D and the FTM-400XD came with the cable required for programming and firmware upgrades. The FT5D came with a cable for doing a firmware update, but I was very disappointed to find that you have to buy a different cable to use the ADMS programming software. You can transfer the data using an SD card, so I decided that the programming cable was not worth the additional expense.

I like the PDN Wires-X mode where one radio acts like a hotspot passing the data through the internet to the Wires-X network, and a second radio is used to access the system. But you must admit that it is a very expensive way to create a hotspot. One advantage of using a radio such as the FTM-400XD in this role is that it is a 50-watt radio. It can cover a very wide area if you connect it to an outside antenna. Other local hams could use it as an access point for Wires-X. Although it receives on one frequency and transmits on a different frequency, it does not act as a duplex repeater, so there should be no licencing issues.

I believe that the search function works better on Wires-X than it does when connected to the YSF and FCS reflectors. It may just be subjective, but it seems smoother to me. You can't use the PDN mode or the HRI-200 without having the Wires-X software running on your PC. It takes up a lot of screen space, but you can run it minimised. The software works quite well other than not having a search function. Most of the Wires-X rooms are regional, but I did notice the N5UDE room. As I write this there are 1703 active Wires-X rooms to choose from. But only 3% of them have ten or more people logged in, and only 1% of them have twenty or more callsigns logged in.

It is generally accepted that **DMR** is the cheapest digital voice option because you can buy cheap Chinese-made radios. But it is much more difficult to set up and more difficult to use. DMR has the steepest learning curve, however, it is the most popular digital voice technology, and it has the most talk groups. You need good computer skills to set up your own code plug, but you may be able to copy an existing one. Adding another channel using the front panel keypad, for instance, if you visit another town and want to use the local DMR repeater, is quite difficult. You have to add a new channel for every talk group you wish to use on the new repeater. It is easier if you have a computer and the programming software handy, but still much more involved than doing the same job on a D-Star or YSF radio.

D-Star is more expensive than DMR, as you are pretty much tied to using Icom radios. With most repeaters and all MMDVM hotspots, any D-Star radio can access any D-Star reflector or gateway. D-Star is fairly easy to operate although there are some rules about how to access and use the reflectors.

WHICH IS BEST?

The big differences are the price you pay for the radio equipment and the complexity of setting up the radios. Each option has 'pros' and 'cons.' Once you have made a purchase and set up your radios, I don't think it matters which digital voice technology you chose. There is no clear winner. The three most popular digital voice modes have similar quality and performance. Many users will tell you that DMR is better because you can use cheap radios, that they prefer D-Star, or that YSF is the easiest to use. It is all very subjective. Some enthusiasts end up with radios for two, or all three, systems. Many D-Star reflectors, Yaesu 'rooms,' and DMR talk groups are linked together, so no matter which system you choose, you can very easily talk to stations that are using a different digital voice mode. A Hotspot can be used with any of the digital voice radios, and even transcoding between systems.

Most D-Star users use the mode because they purchased an Icom radio that includes D-Star. It is easy to use and easy to setup. The mode supports transmitting data files and pictures as well as digital voice. A message which includes the operator's name and often their radio type or location is sent with each transmission. Most D-Star radios include a GPS receiver or the ability to connect an external GPS receiver (IC-9700 and IC-7100). Messages from D-Star radios include the location, bearing, and distance to the station you are hearing. You will be able to access any D-Star reflector worldwide without having them pre-programmed as individual channels on your radio. If you want to talk over a linked repeater or reflector, you must register your callsign with the D-Star organisation. Despite appearing in all of the Icom brochures and manuals, I was never able to get the D-Star callsign and gateway routing options to work. Linking works fine. I have been told that these 'routed' functions only work over genuine Icom repeaters and node connections. D-Star has the advantage that there is only one D-Star network. Both DMR and YSF have competing networks with limited interconnections, which splits the users into groups.

DMR has the advantage and disadvantage that it is a commercial digital voice standard. On the positive side, it means that there are many radios available, ranging from expensive commercial models from Hytera and Motorola to cheaper models from Radioddity, TYT, AnyTone and many other suppliers, mostly from China. The downside is that commercial radios are designed for setup by a radio network operator or a supplier. They are difficult to program, requiring a channel for every talk group on every RF channel. Once programmed they work just as well as the other radios. You can use a local DMR or multi-digital-mode repeater to access a selection of talk groups. With a hotspot and your radio, you can join a local DMR network or several worldwide DMR networks to access a huge number of talk groups.

Unlike YSF and D-Star, the radio must be programmed with the talk groups that you wish to use, not just the RF channels. Being a commercial rather than an amateur radio system, your callsign is not transmitted with each call. To display a name and callsign, you must download a user database into the radio, so that it can match the calling station's DMR number with the stored callsign and name. Most radios can't hold enough callsign information for the currently 228,699 DMR users, and the database grows bigger as more users are registered. You cannot use amateur radio DMR without registering for a DMR ID number.

TIP: Although all DMR radios support analogue FM it is not very good on the cheaper radios. They usually have a very wide front end, leaving them susceptible to interference. The FM audio quality and levels, both transmitting and receiving, is often very low and "tinny." Don't buy a DMR radio if you want to use it primarily for FM. The Icom, Kenwood, and Yaesu radios used for D-Star and YSF are much better quality, and they work fine on FM.

System Fusion is generally considered to be the most expensive of the three modes, but the easiest to configure and use. The expense factor has been alleviated by the release of the Yaesu FT-70D handheld which retails at a very reasonable US $175. YSF is a Yaesu standard. You must use Yaesu radios. One excellent feature is that the Yaesu repeaters support both FM and C4M digital. If a call comes in on FM it is transmitted on FM and the YSF radio will automatically switch to FM to receive it. If a digital call comes in it is repeated as a digital signal. This means that FM and C4M stations can use the same repeater, and C4M stations can switch seamlessly between digital and FM conversations.

BUYING A YAESU DIGITAL RADIO

Here are some things to consider.

- Desktop, handheld, or mobile. What suits your operating style and budget?
    - Desktop radios: FT-991A, FT-991
    - Mobile radios: FTM-100D, FTM-200D, FTM-300D, FTM-400D/XD
    - Handheld radios: FT-70D, FT5D, FT1D/XD, FT2D, FT3D

- Dual band UHF/VHF. All Yaesu C4FM radios are at least dual banders.

- A colour screen that is easy to read and provides plenty of information can be a real bonus. The FT5D and FTM-440XD have colour touchscreens although the touchscreen is not fully implemented on the FT5D. You can make some selections but then have to resort to button pressing. The FT-70D does not have a colour display.

- Waterproof, IPx7 rating or similar if you are planning on hiking or mountain climbing. IPx7 means the radio is protected against temporary immersion in 1m (3.3 ft) of water for up to 30 minutes. The FT5DR is rated at IPx7. The manual states that it is OK for rain or splashes, but that it should NOT be immersed in water. The FT-70 is rated at IPx54 (dust and splashproof). The FTM-400XD is not waterproof because it is a mobile radio. If it gets wet, you have bigger problems.

- Built-in GPS receiver. (FT5D, FTM-400XD, and earlier models)

- Bluetooth can be handy if you are operation mobile or portable. You can connect a headset and keep both hands free for driving or climbing. The FT5D has Bluetooth built in. There is a BU-2 Bluetooth plug-in module available for the FTM-400XD.

- A spare battery, or the ability to connect an external battery, can be very important if you are hiking or activating a SOTA peak.

- If you are planning on using the HRI-200 interface, you need to buy a compatible 'donor' radio. They are the; FTM-400XD (R&E), FTM-400D (R&E) FTM-300D (R&E), and the FTM-100D (R&E).

MYTHS

I often read or hear claims that one digital voice mode has better voice quality than the others. I don't believe that there is any significant difference. Talking through a repeater directly to another station, I would be surprised if you can tell them apart.

You do often hear thready thin audio and sometimes garbled or broken speech, but these are artefacts of the internet data transfer and often transcoding between digital voice data formats. As soon as any digital voice mode is passed through the internet to another repeater, or a talk group, room, or reflector, the compression and packet loss in the internet path will contribute more distortion to the signal than any perceived difference between the Vocoder standards or modulation type being used. This sort of problem will affect all the digital voice modes equally. The Wires-X network may sound better because it is a completely Yaesu-based system. It does rely on the internet for data transport but no transcoding between digital voice platforms takes place.

YSF COVERAGE COMPARED TO FM ANALOG COVERAGE

Like all radio transmissions on the 2m or 70cm band, you can expect 'line of sight' communications when you are operating outside. If you can see the hill or building the repeater is on, you should be able to use the repeater. The situation is different when you are inside a building, in a built-up area, or in woodland etc. Again, you can expect the coverage of a Yaesu C4FM repeater to be similar to an FM repeater.

While the received signal needs to be a bit higher for digital reception, the received speech quality should be much better. On the fringes, YSF generally has superior performance to analogue FM because the forward error correction used in the AMBE+2 digital voice CODEC can cope with bit error rates as high as 5% with no degradation in the perceived speech quality. Digital quality is usually better than FM when operating mobile, especially when the received signal is fairly weak. FM suffers from 'flutter' caused by multi-path propagation.

YAESU RADIOS COMPARED

I bought three Yaesu radios to evaluate while writing this book. They each have good and bad features. Overall, they are top-quality radios.

FT5D

I like the FT5D most of the three radios I purchased. The colour touchscreen is excellent. Weirdly one or two menu procedures start with touch controls but end with an **F Menu** button press. The search function seemed to work better than the search in the FTM-400XD. My main criticism is that you need a different cable for programming with the ADMS-14 software. Given that the supplied cable works fine for transferring firmware to the radio. Why can't it also be used for programming?

FT-70D

The FT-70D is very reasonably priced and it works very well. I know that several YouTubers have said that it is their "go-to" radio in preference to other YSF radios they own. And I can see why. It is smaller than the FT5D and if you already have a reflector connected via your hotspot or a Wires-X room via a donor radio, it is very easy to pick up and use the FT-70D. The numeric keyboard is a big help. There is no volume knob. You have to hold down a button on the side of the radio and adjust the volume with the **Dial** knob. It is not as big a problem as I thought it might be. The biggest issues are the lack of a 'search' function for finding rooms and reflectors, and there are only five 'category' memories for storing the reflectors you visit often.

The FT-70D is an ideal radio for use as an access radio in conjunction with a Wires-X node or a hotspot. You can select the room or reflector using the hotspot dashboard or Wires-X software. It is OK for use in to a local repeater, provided you don't like to 'channel surf' the reflectors and rooms.

I like that it uses the supplied USB cable for firmware updates and programming with the ADMS-10 software. But the ridiculous hit-and-miss procedure that you have to follow to make the programming software work is old-fashioned and, in my opinion, unacceptable. I have programmed radios that cost less than $100 that had far better USB connectivity.

FTM-400XD

The FTM-400XD is a mobile radio with a maximum output power of 50 watts. The XD version has an improved GPS receiver and different firmware, but it is functionally the same as the FTM-400D. I found the search function less effective than the FT5D. Also, it will not trigger my hotspot immediately after turn-on. I don't know why. It is an expensive radio to use as a 'doner' tethered radio in the PDN or HRI modes. The head unit is separated from the radio by a 2m control cable. But the microphone plugs into the main body of the radio, so you can't mount it remotely, at the rear of your vehicle or deep within the dash. I thought that was a bit strange.

Fusion Confusion

Firstly, it is not clear what Yaesu calls their digital voice mode. The boxes my three radios came in all say C4FM which is the modulation method that Yaesu uses. The boxes for the FT-70D and FTM-400XD also state FDMA which stands for frequency division multiple access, implying that the radios carry multiple voice or data signals separated by frequency. C4FM does not carry multiple voice signals like DMR does, but like all digital voice modes, there is forward error correction and signalling data interleaved with the digitised voice signal. Only the FTM-400XD box mentions System Fusion and there is a small image on the back of the box labelled Wires-X, but no mention of it on the front. The Yaesu website says the radios are "Yaesu System Fusion transceivers," and certainly most users and online commentators refer to them that way. There is no mention of Wires-X on the website entries for the FT-70D or the FTM-400XD. It seems that this is no longer a dominant selling point. The manuals that ship with the radios don't say very much about Wires-X either. Just half a page in the FT5D manual, and seven lines in the FTM-400XD manual.

Yaesu System Fusion, commonly known as YSF, or just 'Fusion,' refers to the way that Yaesu DR-1, DR-1X, and DR-2X repeaters manage digital C4FM and analogue FM. If an analogue call is received, it is re-transmitted as analogue FM. C4FM (YSF) radios will automatically switch from digital to analogue to receive it, and of course, FM radios will be able to receive it as well. If a digital C4FM signal is received by the repeater it will be regenerated and re-transmitted by the repeater as a digital signal. In this case, only C4FM-capable radios will be able to hear the transmission. There is also a transitional mode where the repeater will receive either C4FM digital or analogue FM signals but re-transmit all signals as FM. This means everyone can hear either analogue or digital calls, but the repeater is not acting as a digital repeater, so the benefits of using digital transmission are lost. As far as I know, this feature of the repeater acting as both an analogue FM and digital voice repeater is unique. Especially the digital input to analogue output mode. It is great because you only need one repeater on the hilltop to provide service to FM and digital voice operators.

Wires-X

Wires-X, the 'Wide coverage Internet Repeater Enhancement System' is a Yaesu-owned and operated internet-based reflector system. It is accessed using the Wires-X button or a couple of keystrokes on your Yaesu radio, through either a genuine Yaesu repeater or a Wires-X PDN (portable digital node). A PDN is a Yaesu radio that is connected via a data cable through a PC to the internet. In this mode, the radio does not transmit. It is similar to the D-Star Terminal mode. The third possibility, which allows you to host a Wires-X 'room' is to setup your own Wires-X node. This requires a compatible Yaesu radio, plus a Yaesu HRI-200 interface and an internet connection. See https://www.yaesu.com/jp/en/wires-x/id/active_room.php for a list of active Wires-X rooms.

Wires-X should not be confused with reaching YSF reflectors via a hotspot or a non-Yaesu repeater. Even though you access that network by pressing the same Wires-X button or keystrokes on your radio. It is quite confusing.

YSF reflectors

Using the YSF (Yaesu System Fusion) network is very similar to the way that other radio amateurs use DMR, D-Star, NXDN, and P52 talk groups and reflectors. You use your Yaesu radio to trigger a hotspot, which receives your transmission and acts as a modem, sending your call over the internet to a YSF reflector. You can make calls to users of the YSF reflector, any connected repeaters, or other linked reflectors. A few YSF reflectors are linked to Yaesu Wires-X rooms, but most are not. Many are linked via XLX reflectors to other digital voice mode, reflectors and talk groups. In this way, you can use your YSF radio to talk to amateur radio operators who are using D-Star, DMR or other radios. They might not be using any kind of radio. They could be talking via an Android phone app and a DV dongle, or a computer program.

Non-Yaesu and 'multi-DV mode' repeaters typically offer local coverage and access to YSF reflectors over the internet. Simplex repeaters or high-power hotspots offer internet-linked connections, but they don't act as a local repeater. You cannot directly access YSF reflectors via the Wires-X network, but Wires-X users can talk to hotspot users via the interconnected YSF to Wires-X rooms.

YSF Reflector ID numbers can be found at: https://register.ysfreflector.de/ The list includes links to many of the reflector dashboards. They are also listed at https://www.pistar.uk/ysf_reflectors.php

FCS reflectors

The FCS (Fusion Connect System) is a centrally located System Fusion reflector service. There are currently 24 FCS servers with 100 modules each. 1122 of the 2400 reflector modules are allocated. The rest are currently not used. Module 99 is used for the Echo function on eight of the servers.

The FCS reflectors are similar in concept to the YSF reflectors except YSF reflectors are de-centralised and hosted by individual hams.

If you are using a hotspot, you can select an FCS reflector from the admin page on the Pi-Star dashboard. From the radio, you can enter the reflector name or the FCS reflector code into the search function. For example, **FCS00290** is America Link. America Link is a Wires-X Network room that is bridged to FCS00290 and FCS0039.

The reflectors can be saved into one of the five 'category' memory banks by selecting **Add** immediately after the node connects. This seems to be the only time you can do this. There is a list of the FCS reflectors at https://www.pistar.uk/fcs_reflectors.php. Also, see http://xreflector.net/ for the reflector dashboards.

YCS reflectors

YCS is the latest reflector system that can be accessed with a Yaesu C4FM radio. It is a multi-protocol server capable of handling YSF, FCS, and IMRS, which is a newer linking protocol supported in the Yaesu DR-2X repeater (but not the older DR-1X). It means you can set up one reflector server that does everything! YCS reflectors can connect to Yaesu repeaters, non-Yaesu repeaters, and hotspots. There are currently 20 YCS servers with up to 38 active modules. You can link to a YCS module using the FCS network. For example, YCS262 module 20 is accessed by linking to FCS26220.

One of the biggest advantages of DMR over D-Star and YSF is that it can be linked to several static talk groups at the same time. On the DMR+ network, your repeater can be linked to a selection of talk groups, or you can configure your hotspot to listen for a range of talk groups. On the Brandmeister network, you can select static talk groups for TS1 and TS2 on the excellent Brandmeister dashboard. Normally YSF and D-Star can only be linked to a single reflector or reflector module. That module can be connected to several other talk groups and reflectors, but it is not the same. Wouldn't it be great if you could listen for calls on America Link, Japan Link, and CQ UK at the same time? The biggest advantage of the YCS servers is that they are designed to work like DMR. You can configure your hotspot, or BlueDV, to listen for several modules on the same YCS server. This 'DG-ID mode; is not supported by the 'standard' version of Pi-Star available from https://www.pistar.uk/. But, there is a 'fork' created by Manuel Sanchez Raya EA7EE http://pistar.c4fm.es/ which does support the YCS DG-ID mode. Download it at,
 https://drive.google.com/file/d/1jU6Bia-DDkdH4bePjGDn_Iph6qMdOrts/view

A good place to start with YCS reflectors is to click the **DVMatrix** YCS link in the left panel at http://xreflector.net/. The 'Activity' tab shows network calls on each of the 20 YCS servers. A green bar indicates a transmitting station. Amber bars indicate linked stations that are receiving the transmission. Blue indicates that the station has finished talking and a countdown timer is operating. The 'Group' numbers show which modules are active on each server. The 'DG-ID List' shows the nationality or use, and language used on each module. For example,

- YCS530 module 35 is linked to YSF ID41562, 001-CQ-UK   (FCS53035)
- YCS530 module 37 is linked to YSF ID32592, America Link   (FCS53037)
- YCS530 module 22 is linked to YCS222 Italy   (FCS53022)
- YCS530 module 53 is linked to YCS530 New Zealand   (FCS53053)

YCS was developed to support the various protocols that are available for the exchange of C4FM streams. The new System Fusion II IMRS (Internet-linked Multi-site Repeater System) protocol which is available for the new DR-2X YAESU-Repeater is also supported. This makes it possible to combine Repeaters and HOTSPOTS in one server.
http://ycs-wiki.xreflector.net/doku.php There is more about YCS at…
http://ycs-wiki.xreflector.net/doku.php?id=ycsserver:overview:server#ycs_server_overview

YCS can handle data from Yaesu repeaters and hotspots. The dashboards show who is transmitting and all the other nodes and callsigns that are hearing the call. The design is very similar to the DMR IPSC dashboards.

If you click the **YCS530 (NZ)** YCS link in the left panel at http://xreflector.net/. You will see the YCS530 dashboard. On the top line, there is the link to DVMatrix which brings in modules from other YCS servers. Then there are eight highlighted links which are permanent links to popular YSF reflectors. The remaining lines indicate nodes and individuals which are linked to YCS530. The number in the DG-ID column indicates the module they have selected. For example, the ones with 37 are listening to America Link because that's what module 37 is connected to. The 'Interlink' tab shows five links to IPSC2-NZ-HOTSPOT which is one of the New Zealand DMR servers. Each of those links connects a DMR talk group on NZ-HOTSPOT to a module on YCS530. This allows Yaesu users access to those four DMR talk groups.

YCS on BlueDV

You can use the YCS system by linking to a connected FCS reflector. For example, this connection to FCS530 module 37, America Link on YCS530.

If you create a link and transmit for a few seconds, you should see your callsign come up on the YCS dashboard. Find the correct dashboard in the left panel at http://xreflector.net/. Since we connected to YCS530 the dashboard is at http://ycs530.xreflector.net/ycs/#.

If you want to operate the unique arrangement where you will hear calls on several modules, you need to make an easy change to the **BlueDVconfig.ini** file. See page 67 for setup instructions.

Nr.	Repeater	Name	QRG	ID	DG-ID
26	FCS53037 -ZL3DW (05)	Hamshack	434.3000	4028	05 35 37 80 84 89

The image above shows my callsign on the YCS530 dashboard along with the six static DG-ID modules I have selected. With this setup, I will hear a call on any of the selected modules. If a call is in progress, I will not hear the other modules until the channel has been clear for 20 seconds. These 'static' modules will be linked whenever BlueDV is connected to any of the 20 YCS servers. Looking at the picture, you can see I am linked to FCS530 module 37, which is operating on a YCS server. I'm not sure where the name 'Hamshack' came from. The QRG is the frequency I entered into the BlueDV setup. It is meaningless because I am not using a radio connected via the computer's serial port. I think that the four-digit ID number is allocated by the server. If it was a DMR dashboard the ID column would contain your DMR ID number.

WORLDLINK

WORLDLINK YSF00008 is technically a YSF reflector. It was established by David PA7LIM the creator of Peanut and BlueDV. The difference between YSF00008 and a normal YSF reflector is that it has modules, or 'rooms,' like an FCS reflector, and it has a dashboard at http://worldlink.pa7lim.nl/. David Grootendorst calls them 'Treehouse' reflectors. https://www.pa7lim.nl/treehouse/. The EA7EE version of Pi-Star supports the selection of static rooms on both Worldlink and Europelink.

You can change rooms by changing the DG-ID number on your transceiver and transmitting a 'kerchunk.' For example, setting a TX DG-ID of 01 puts your transmission into the 'Parrot' room. Setting a DG-ID of 19 puts your transmission into the 'CQ - Canada' room.

It is probably easier to leave your DG-ID at TX 0, RX 0, connect to the reflector, and use the 'room control 'dropdown list on the reflector dashboard to change rooms.

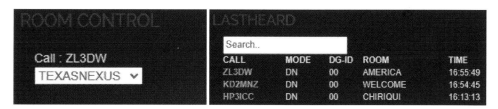

Figure 1: Worldlink room selection

EUROPELINK

EUROPELINK YSF00007 is another YSF reflector 'with rooms.' It was established by David PA7LIM for European users. It has a dashboard at http://europelink.pa7lim.nl/.

The rooms are not all based in Europe. There are several Central and South American rooms, and links to CQ-UK, DVScotland, and GB Hubnet.

The dashboard has the same room control dropdown list and tables showing 'last heard, 'online stations,' and 'rooms,' as the Worldlink site.

OTHER PA7LIM SERVERS

David is developing similar 'Treehouse' servers for DMR at http://dmr.pa7lim.nl/ and D-Star at http://dcs018.xreflector.net/.

PEANUT ROOMS

Peanut has links to rooms, reflectors, and talk groups for all of the digital voice modes, plus several Peanut rooms that are for Peanut users only. It does not provide access to all rooms on all networks, but you can reach the most popular ones.

There is a list of Peanut rooms at http://peanut.pa7lim.nl/rooms.html.

Yaesu digital radio

You can operate your Yaesu digital radio in several ways.

1. Simply selecting DN (digital narrow) means that the radio will transmit and receive using C4FM digital modulation. You can use this mode to talk directly to another radio in simplex mode, to local hams via a Fusion or non-Yaesu repeater, or worldwide via internet linking. The DN mode is preferred for hotspot, linked repeater, or Wires-X operation.

2. The VW (voice-wide) mode is good if you are using simplex or a local repeater. The audio quality is greatly improved because the full 12.5 kHz bandwidth is used for the voice signal. However, the signal strength needs to be a little stronger because there is no forward error correction.

3. The DW (digital-wide) mode is used for sending data such as photographs at a higher speed than the DN mode. It is a data-only mode, with no voice.

4. The FM mode is good old analogue FM.

TIP: A bar above, $\overline{DN}$, $\overline{VW}$, $\overline{DW}$, or $\overline{FM}$ means that the radio is in AMS (automatic mode select) mode, in which it will change the mode to suit the received signal. In some cases, this is annoying as the radio will switch to FM and the squelch will open if an interfering signal is received on the frequency.

Digital simplex

You can call another radio directly without the aid of a repeater or hotspot. Simply select a digital simplex frequency (check your local band plan). Select the DN or VW mode and call the other radio. Both radios in the conversation must be in the same digital voice mode or set to AMS.

Local Repeater

You can make local calls via a CF4M or multi-mode digital repeater, using either the DN mode or the VW mode. Both radios in the conversation must be in the same mode or set to AMS. I suggest that you start in the DN mode and switch to the VW mode by agreement.

Set your radio to the repeater output frequency. The transmitter offset should be selected automatically, but if it is not, then you can set it in the **Config** part of the main menu. On an FTM-400XD or FT5D, it is **RPT Shift** and **RPT Shift Freq.** On an FT-70D long-press the **F** key and select **46 RPTFREQ**. Set the offset, then PTT to exit the menu. **F > 0 > Dial** changes the repeater shift. Make sure that you set positive of negative offset according to the band plan. Use the D/X key, GM/X. key or Mode button according to your radio to set the mode to DN, VW, $\overline{DN}$, or $\overline{VW}$.

Wires-X

You can access the Yaesu Wires-X network to make calls worldwide. It also has a local and international news function and short messaging rather like texting on your phone. To access Wires-X you must transmit through a genuine Yaesu repeater that has internet connectivity, use a compatible Yaesu radio in the PDN/HRI mode, or access a Wires-X node via a Yaesu radio connected to a Yaesu HRI-200 interface and the internet. Set your radio to the repeater output frequency, offset, and DN mode as discussed in 'local repeater' above.

Long press the **D/X** button (FTM-400XD), **GM/X** button (FT5D), or **F > AMS** on an FT-70D. If your signal reaches the repeater and nobody is currently using it, the radio should connect. If the connection is refused, the repeater may be busy, or it might be out of range. Try again in a minute. If the repeater is a genuine Yaesu repeater connected to Wires-X you will be connected to Wires-X. If the repeater is not a Yaesu one or it is connected to the YCS network, you will connect to the public network. There may be a reflector connected or there might not be.

DIGITAL GROUP ID (DG-ID) AND GROUP MONITOR (GM)

The digital **Group Monitor (GM)** mode is accessed using the **GM** key on your radio. It will poll the repeater every few seconds to find any stations that are within its coverage area that are using the same digital ID code. If your radio is set for a digital ID of 00 it will find all stations that are within the repeater coverage area. The stations in the group will be listed on the display, along with their distance and bearing if the radios are equipped with GPS receivers.

TIP: the GM button is also used for selecting a module when you are connected to a YCS reflector. There is more about that in the 'Using a YCS reflector' chapter.

A two-digit **'DG- ID'** (digital group identification) code is transmitted along with the voice and data transmission. It works like a squelch system. The DG-ID numbers for receiving and transmitting are usually set to the same number, normally 00. Setting a receive ID of 00 is a bit like open squelch. You will hear all callers using the repeater. If you set any other number, you will only hear callers that are transmitting that specific digital ID code.

Some C4FM repeaters use a digital ID number as an access squelch in the same way that an FM repeater might require you to transmit a CTCSS tone. In that case, you would set the repeater code as your transmitting digital ID. Presumably, this would be done to limit access to a group of people rather than to combat interference, since the C4FM demodulator would reject anything other than a genuine C4FM signal. It creates an odd situation where only stations that are transmitting the correct code will be able to open the repeater squelch, but any station using 00 as the receiving digital ID code will be able to hear the signals on the repeater output.

In the drawing below, radio 1 and radio 2 are in a group that is using DG-ID code 50. They can hear each other, and the group monitor (GM) function will list the two radios in group 50. Neither radio will be able to hear radio 3 or radio 4 when they use the repeater.

Radio 3 is set to DG-ID code 00. Its owner will hear any of the other stations but none of the other stations will be able to hear them. This is a confusing situation for the owner of radio number 3. The GM function will not list any other stations because none of the other stations are transmitting code 00.

Radio 4 is in a different group. It is set to DG-ID code 20. Its owner will not hear any of the other stations because none of them are transmitting with code 20. The GM function will not list any other stations because no stations in the 20 group are within range of the repeater.

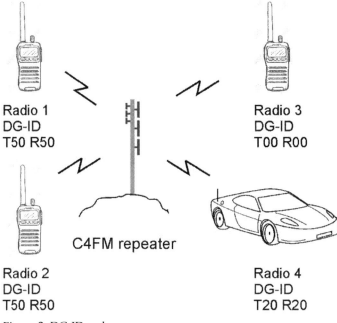

Figure 2: DG-ID codes

If for some reason the repeater is set for a receive DG-ID code of 50, only radio 1 and radio 2 would be able to use the repeater, although radio 3 would be able to hear them using it. Radio 4 would not hear or be able to trigger the repeater. Setting the repeater to transmit DG-ID code 50 as well, would make no difference.

Nobody would do this! Leaving the repeater receiver ID code at 00 and setting the transmitter code to 50 would allow radio 4 to trigger the repeater but not hear the repeater output. In this situation, any C4FM radio would be able to trigger the repeater but only those with their receive DG-ID code set to 00 or 50 would be able to hear it.

ROOMS & REFLECTORS

Rooms and reflectors extend your access outside the area covered by your local repeater, allowing you to talk with amateur radio operators all over the world. With Yaesu radios, there are two ways that this can be accomplished. You can use the Yaesu-owned and operated Wires-X network, or the YSF, FCS and YCS networks with their cross-connects into Wires-X rooms, reflectors, and the talk groups and reflectors of other digital voice networks.

A 'room' is the Wires-X equivalent of a DMR talk group or a D-Star reflector. Of the three terms, I think that 'room' best describes the function. You can think of it as a repeater that is hosted on the internet or a meeting place where like-minded people hang out. To minimise confusion, I have elected to refer to Wires-X servers as 'rooms' and YSF, FCS, or YCS servers accessed via a hotspot or a non-Yaesu repeater as 'reflectors.' Some people call those 'rooms' as well.

Getting into a 'room' is a little different in the Yaesu system. You do not have to have every channel programmed with talk groups the way you do with DMR, and there is no requirement to register with a network operator.

You cannot choose which other repeaters and user 'nodes' are connected to a room or a reflector, but you can choose the one you want to connect to. Users within each repeater coverage area can connect or disconnect their repeater as they wish. It is good etiquette to ask before you disconnect your local repeater from a reflector. Someone might be waiting for a call or for a net to begin. Since you won't affect anyone else, you can connect or disconnect your hotspot as often as you like. Once the link has been established you can put out a CQ call or respond to others on the channel. Any call you make to another station is not private. It will be broadcast on the room or reflector, any other connected reflectors, and possibly repeaters attached to those.

The rules if you are using a repeater, are the same as farm gates. "Leave them the way you found them." If you established a connection to a room or reflector, you should send the 'unlink' signal to remove the link after you have finished making calls. If possible, you should probably also re-instate any link that was in effect before you started. If you didn't establish the connection and you don't want to make a different connection, you should leave the link connected. The rules may be different for your local repeaters.

If you are using the Wires-X software in PDN, HRI, or HRI-200 mode, or a hotspot or DV dongle, you can link or unlink rooms and reflectors any way you want, as often as you want. Leaving a room or reflector connected when your radio is unattended may mean that you miss someone calling you after seeing your callsign listed. Pi-Star hotspots reset every night and reconnect to the nominated reflector. So, many of the people listed on a dashboard, may not be listening.

GATEWAYS

Gateway operation is available on Wires-X, but it is not commonly used. A gateway link is when you connect directly to a repeater or a node like a hotspot, rather than linking to a room or a reflector. I have borrowed the term from D-Star, Yaesu calls it 'connecting to a node.'

Not all nodes on the Wires-X node list can be reached. The ones that can are connected using the Wires-X software, or from your radio by searching on the node name, usually a callsign. Or by entering their node ID number, listed as DTMF on the node list. Use **Search & Direct > Search & Direct > ID** and enter the number. Or **Search & Direct > Search & Direct** and enter the start of the node name. The first four or five characters are usually enough. Unfortunately, it is not a context-sensitive search so you cannot enter the text from the middle of a node or room name.

As far as I know, you cannot connect to an individual callsign or repeater via a hotspot or repeater linked to the YSF, FCS, or YCS network.

CATEGORY MEMORY BANKS

The FT5D and FTM-400XD have five banks of 20 memory slots specifically for storing rooms and reflectors. You can name each bank and add your favourites. On my FT5D, I have 'YSF,' 'FCS,' and 'Busy,' for when I am using my hotspot. And 'Wires-X' for when I am connected to the Wires-X network.

*TIP: When you select the **Search (white arrow)** on an FTM-400XD, or **Search & Direct on an FT5D**, and **C1**: to **C5**: the radio will poll the network to check that the rooms or reflectors in the list are active. Only active rooms or reflectors will be displayed. So, if you select the Wires-X list when using a hotspot, only Wires-X rooms that are linked to the YSF or FCS network will be displayed. If a category bank is empty, it will appear greyed out and it cannot be selected.*

FTM-400XD category memory slots

Unfortunately, the FTM-400XD only supports adding Wires-X rooms to the category memory banks. It will not let you add reflectors. The ADMS-7 software does not have an edit function for the category memory banks, so you can't add them there either.

You may notice that the radio transmits immediately after you have connected to a new room. The Wires-X node responds to this by displaying the room's information screen. The one that says 'Emergency,' News Station,' 'Local News,' 'Int News.'

At the bottom of that screen is the ADD button. If you want to add the selected reflector to the category list, touch **ADD** then the **C1** to **C5** category list that you want to use.

To select a previously saved Wires-X room from the category list. Press the **DX** button to link to a Fusion repeater or Wires-X node. Touch the white search arrow then select **C1** to **C5**. You may have to touch it twice. The radio will transmit, and the saved list should appear. Touch the **room** that you want and if it is not busy, you should be connected.

To name the category banks:

- Long press the **Disp/Setup** button to get to the main menu.

- Select **Wires** > **Edit Category Tag** > touch a **C1** to **C5** label >

- Type in a **name** up to 16 characters, then touch **ENT**.

- Long press the **Disp/Setup** button to get back to the Wires-X mode.

FT5D category memory slots

Adding reflectors or rooms to the category banks usually works on the FT5D, but you can only do it immediately after connecting to the room or reflector. It might not work on all reflectors. Follow this process exactly.

- If not already there, press and hold **GM/X** to enter the connected mode

- Press and hold **Band** to unlink. You might not have to do this to change to a different reflector or room, but you usually do, if you want to add to the category memories.

- Touch **Search & Direct**. Then use **Search & Direct** or **ALL** to find the reflector or room you want to add to a category memory slot. Touch the entry to connect to the reflector or room.

- The 'news and information page' should come up. Touch **ADD** at the bottom of the screen. As far as I know, this is the only way to get to the ADD option. Then touch the **category bank** that you want to use. The reflector or room will be added to the bottom of the list.

- **Back** > **Back**

Selecting a reflector or room from the category list is easy. In the connected mode, touch **Search & Direct** > **category bank** > **reflector**. You can use the **Dial** knob to select the category bank and the reflector.

To name or edit the category banks press and hold **F Menu** > **Wires-X** > **Edit category tag** > **C1** to **C5** (select with F Menu key) > **edit the title** > **Back 4 times**.

To delete a reflector or room entry, press and hold **F Menu** > **Wires-X** > **Remove Room/node** > **C1** to **C5** > **select the item to be removed** > **Back 4 times**.

TIP: the link information displayed on the radio can be incorrect if you used Pi-Star or a different radio to connect to a reflector. It should eventually update, but it can be confusing if the radio's display says it is linked to a different reflector. The link shown on the Pi-Star dashboard is the 'real deal.'

FT-70D category memories

Set the radio to the 'connected' or 'Wires-X' mode using F > AMS.

The FT-70D only has five programmable category memories, C1 to C5. They can be used to store reflectors, or rooms if you are using Wires-X. To save a link in one of the C memories, simply connect to the room or reflector and hold down the corresponding number key 1 to 5 on the keypad until you hear a beep.

Cn 'Connected node' displays the currently connected node

C0 displays the most recently connected node

C1 to C5 'Category memories' displays saved reflectors or room links

Lc displays the callsign of a station being received, or your repeater or hotspot when the channel is idle

En * can be used to enter the DTMF number of a room or reflector

TIP: When you are connected to an FCS reflector, Lc displays the connected reflector name, C0 shows the FCS reflector number and Cn shows the module letter.

Select a previously saved room or reflector by turning the Dial to show the category, C1 to C5. Press AMS to connect. If it does not connect, the reflector or room may be busy. Try again after a few seconds.

If you want to select a different room or reflector turn the Dial to show En *. If you are using the public network, enter the YSF DTMF ID (not the YSF number), from the list at https://register.ysfreflector.de/. If you are using a Wires-X node enter the Wires-X room ID. Obviously, you cannot reach most Wires-X rooms unless you are connected to a Wires-X node.

You can't "officially" use this method to connect to FCS reflectors. The most reliable method to do that is to create the link using the dropdown box on the admin page of your Pi-Star dashboard. You can sometimes enter the reflector's FCS number into the En * field, but the FCS numbers don't usually align with the required DTMF codes. The link often does not connect, and incorrect codes can cause Pi-Star to disconnect from the network.

Using the En * field to enter an FCS reflector number works much better on the EA7EE version of Pi-Star, which interprets the FCS number and makes the correct network connection. Check out the section about this on page 43.

RADIO MEMORY BANKS

The radios have memory banks that can hold digital voice or FM channels.

FT-70D memory channels

The FT-70D has 900 memory channels which can be split into 24 memory banks. Unusually, a memory channel can be associated with more than one memory bank. For example, a channel could be saved in '70cm,' and 'local repeaters,' and 'DV.' There are also 50 pairs of frequencies in the PMS bank, which set the lower and upper limits of the programmable scans. You can set a frequency offset into the PMS scan frequencies. I guess it's for scanning repeater output sub-bands. Weirdly you can set the lower frequency to a different offset to the upper frequency. I wonder what happens during the scan.

There are six 'home channels' and 99 'skip channels.' Skip channels are frequencies that will not be scanned. They do not have to be memory channels. You would normally set a skip channel if a frequency had a fixed carrier or an interference signal that is causing your scan to halt. There is a home channel for each band.

You can enter memory channels using the ADMS-10 programming software. Or directly on the radio. To save a frequency into a memory slot, set the frequency, offset and mode. Press and hold the **V/M** key. Rotate the **Dial** knob to select a channel number. Empty channel numbers will blink. Used channel numbers will not. The F icon will also blink. Press the **V/M** key. **M-WRT?** will appear on the display if the channel is already in use. Press **V/M** again to overwrite the channel. Then the memory name tag will appear. Text and numbers can be entered using the keypad or by rotating the Dial. I find the keypad entry easier. **Band** moves the cursor right. **Mode** moves the cursor left. Press and hold **GM** to erase all characters after the current position. Finally, press and hold **V/M** to save the channel.

To recall a memory channel, press the **V/M** button to enter the memory channel mode, and use the **Dial** to change channels. Or enter the three-digit memory number.

It is not a bad idea to set your hotspot or local Fusion repeater to the **'home' channel** for the band. It makes it easy to find. Switch to the channel or frequency, then press and hold the **HM/RV** key. To recall the channel, press **F > HM/RV**.

FT5D memory channels

The FT5D has 900 memory channels which can be split into 24 memory banks. A memory channel can be associated with more than one memory bank. For example, a channel could be saved in '70cm,' and 'local repeaters,' and 'DV.' There are also 50 pairs of frequencies in the PMS bank, which set the lower and upper limits of the programmable scans. You can set a frequency offset into the PMS scan frequencies. It is probably for scanning repeater output sub-bands.

There are eleven 'home channels.' One each for the eleven bands on VFO-A; AM BC, SW, 50 MHz, FM BC, AIR, 144 MHz (2m), VHF, INFO band 1, 430 MHz (70cm), UHF, and INFO band 2. VFO-B can only use six of those bands.

In addition, there are pre-programmed memory banks for the 57 International Marine band channels, 10 weather channels (US version only), and 89 worldwide shortwave broadcast channels.

There are 99 'skip channels.' Skip channels are frequencies that will not be scanned. They do not have to be memory channels. You would normally set a skip channel if a frequency had a fixed carrier or an interference signal that is causing your scan to halt.

You can enter memory channels using the ADMS-14 programming software. Or directly on the radio. To save a frequency into a memory slot, set the frequency, offset and mode. Press and hold the **V/M** key. The next available memory channel is displayed. If you wish, you can rotate the **Dial** knob to select a different channel. A white-blinking channel number indicates a free slot. A red channel number indicates a used memory slot. >> steps 10 channels up and << steps 10 channels back. Press **V/M** to save the channel. If it is in use, you will get an **Overwrite?** message. Touch **OK** to overwrite the memory slot or **Back**, to abort. Input a name 'memory tag' for the channel. Press **V/M** or PTT to exit.

To recall a memory channel, press the **V/M** key to get into the memory channel mode. Rotate the **Dial** knob to select the channel you require.

TIP: short press **F Menu** *and use the* **Dial** *knob to change 10 channels per click. Short press the* **F Menu** *key again to return to single channel steps.*

MAG (memory auto grouping) automatically groups memory channels into bands. The **Dial** knob will only select channels in the selected band. Enter the memory mode with **V/M**. Press the **Band** button to step through the MAG bands and use the **Dial** knob to select a channel. The A-Band can select from ALL, AIR, VHF (2m), UHF (70cm), AM, FM, SW, and OTHER. Note that if there are no saved memory channels in a band it will not be available for selection. The B-Band can select from ALL, AIR, VHF (2m), UHF (70cm), SW, and OTHER. You access frequencies on the 50 MHz band, FM BC band, and UHF2 band, on the Band B VFO.

TIP: very annoyingly unless ALL is selected, the red memory bank identifier will flash. I thought it was an error, but it is normal.

It is not a bad idea to set your hotspot or local Fusion repeater to the **'home' channel** for the band. It makes it easy to find. Switch to the channel or frequency, then press and hold the **V/M** key. Touch the **six white squares icon** > H.Write > OK > OK. Enter a name. If it was a channel, it will already have a name. Press **V/M** or PTT to exit. To recall the home channel, press **F Menu** > **Home**.

FTM-400XD memory channels

The FTM-400XD has 1000 memory channels. 500 for Band A (upper display) and 500 for Band B (lower display). They can be stored on the SD card as a backup. There are nine pairs of frequencies in each PMS bank. They set the lower and upper limits of the programmable scans. The upper and lower displays each have a set of five home channels for the different frequency bands. They are AIR Band, VHF, GR1 which is the 222 MHz band, UHF, and GR2 at 850 MHz.

You can enter memory channels using the ADMS-7 programming software. Or directly on the radio. To save a frequency into a memory slot, set the frequency, offset and mode. Press and hold the **F MW** key. The frequency will be displayed in the next available memory slot. You can use the **Dial** knob if you want to move the frequency to a different memory slot. Perhaps, to overwrite an existing channel. To name the channel, press and hold **F MW** and a keyboard will pop up. Enter the channel name, then **ENT**. Press **F MW** again to save the channel. If you are overwriting an existing channel, say **OK** in response to the 'Overwrite?' question.

To recall a memory channel, touch **V/M** to enter the memory channel mode. The last used memory channel will be displayed. Use the **Dial** knob to switch between occupied memory channels.

To erase a channel, touch **V/M** for three seconds, then release it to enter the memory channel mode. The last used memory channel will be displayed. Use the **Dial** knob to switch between occupied memory channels. Touch **DEL** to delete the channel, and **OK?** to confirm. Touch **Back** to exit.

It is a good idea to set your hotspot or local Fusion repeater to the **'home' channel** for the band. It makes it easy to find. Switch to the channel or frequency, then press and hold the **F MW** key. Use the **Dial** knob to select the HOME memory slot. It is before channel 001 and after P9U. To name the channel, press and hold **F MW** and a keyboard will pop up. Enter the channel name, then ENT. Press **F MW** again to save the channel. Say **OK** in response to the 'Overwrite?' question.

TIP: this can go a bit strange if you are saving a memory channel frequency as the home channel and the VFO is not set on the same band. I recommend that you set the VFO to a frequency in the 70cm band if the memory channel you want to save as a home channel is in the 70cm band or a frequency in the 2m band if the memory channel is in the 2m band.

To recall the home channel, press **F MW** -> (if reqd.) -> (if reqd.) **> Home > F MW.**

TIP: the FTM-400XD does not have a band switch so it might be a good idea to put UHF channels on the top display and VHF channels on the bottom display. This rule can be relaxed if you want to monitor two VHF channels or two UHF channels at once. Digital communications including Wires-X can only be performed on the upper 'Band A' display. FM APRS only works on the lower 'Band B' display.

FT5DE REGION SETTING

There is a procedure for changing the region of the FT5DE to suit the local repeater offset, which is different in the UK compared to Europe. This method only works on the 'European' FT5DE. It does not affect the International FT5DR radios. There are links to two good YouTube videos below, but I will step through the process. I don't know of any similar process for the FT-70DE or the FTM-400XDE, and I don't know any magic codes for the international versions of the radios.

You can change a UK radio to the European B2 setting or an EU radio to the UK C2 setting.

First, you need to save the current radio settings and saved memory channels to an SD card. This is essential because changing this setting does a full reset on the radio.

Make sure you have a micro SD card plugged into the radio. Details on page 78.

Hold F MENU > SD CARD > BACKUP > use the Dial knob to select **Write to SD** > F MENU > OK > OK. When it has been completed, press BACK > BACK.

Select **DISPLAY** > **OPENING MESSAGE** > use the Dial knob to select **MESSAGE** > then F MENU > enter AH082M with the letters in capitals. Press the PTT to exit.

Turn the radio off. Hold down **GM/X** and **V/M** while you turn the radio on.

When the radio re-starts a European radio should display TYPE: >B2 a UK radio should display TYPE: >C2.

If it is incorrect for your region use the **Dial** knob to change the setting and then touch and hold **WRITE** until there is a beep and the radio restarts.

The radio will be reset and will ask for your callsign. You can enter any letters because you are about to recover your settings including your correct callsign from the SD card.

Hold F MENU > SD CARD > BACKUP > use the Dial knob to select **Read from SD** > F MENU > OK > OK. When it has been completed, press BACK > BACK.

The final thing is to stop displaying the message when you start the radio. Select **DISPLAY** > **OPENING MESSAGE** > use the Dial knob to select **DC** then press the PTT to exit. You could enter a personal message in place of the code if you wish.

How to change FT5D from Europe B2 to UK C2 repeater format (M0FXB) https://www.youtube.com/watch?v=EGAeUTZ5ktA

Change the Yaesu FT5DE Region to UK (Yaesu UK) https://www.youtube.com/watch?v=CCBFj1XNmKs

FIRMWARE UPDATES

If you purchased a new radio, you probably won't have to update your radio firmware, but it is worth checking that it is up to date. Firmware updates eliminate bugs that may have been annoying you.

I had problems getting the PC to accept the connection from the radios but did eventually manage to perform the firmware updates without further incident.

FT5D firmware update

The first step is to find out if a firmware update is required. There is no point in performing a potentially risky firmware update unless you have a major problem with the radio, or there is a bug fix update available. To check the current firmware, Hold F MENU > DISPLAY > SOFTWARE VERSION. The item of interest is **Main Version** but take note of the **Sub** and **DSP** versions as well so you can tell if they get updated. Check out the firmware available on the Yaesu website as covered below. You can see that the currently available firmware is version 1.11 issued on the 23rd of March 2022. No update is required if it matches the version on the radio.

Always backup your settings and memory channels before doing a firmware update. Usually, the radio will be reset, and all of your additions will be erased. Hold F MENU > SD CARD > BACKUP > use the **Dial** knob to select **Write to SD**, then press F MENU > OK > OK. When the update has been completed, BACK > BACK.

There is a short note about the firmware in the advanced manual, but no instructions for carrying out a firmware update. Navigate to the Yaesu website, find the FT5D and click on the link. Select the 'Files' tab. There are separate firmware files for each version of the radio. Download the FT5DR or FT5DE firmware file. I downloaded the USA version because I have an FT5DR. There is a different version for Japan which is only available on Yaesu's Japanese language website. You might need the Prolific USB driver. You can download it now or wait until you need it. You could download the update instruction manual. But you won't need it.

FT5DE_EXP_Firmware Ver1.11_Update_2022_03.zip (3/29/22) FT5DE version for EU

FT5DR/DE MAIN/SUB Update Instruction Manual (2204-B) Update manual

FT5DR/DE Update MAIN Firmware Ver1.11 Information (3/29/22) What changed?

FT5DR/FT5DE Programming Software ADMS-14 (Ver.1.0.1.0) ADMS-14 programming

FT5DR_USA_Firmware Ver1.11_Update_2022_03.zip (3/29/22) FT5DR USA/Intl version

PL23XX_Prolific_DriverInstaller (Ver. 402) Prolific USB Driver Installer (Ver. 402)

WIRES-X Connection Cable Kit Driver Installation Manual (2205A) Installation of Wires-X

Unzip the file to a temporary directory on your PC. The 'Downloads' directory is fine. The USB driver installation and instructions are in the program. Do not load the prolific driver or run the HMSEUSBDRIVER file.

Run the **FT5D_MAIN_ver111(xxx).exe** file. If this is the first time you have run the firmware update, click the **USB Driver Install** button. The COM port and Baud Rate boxes are not used. They can be ignored. After the driver has been installed, the **Update** button should be available. If it is not, download and run the prolific USB driver. Then run the Firmware program again, but don't click the USB Driver Install button. Simple?

The radio was supplied with the mini-USB to USB type A cable required for the firmware upgrade. Nothing magic. It is a standard cable. Why it is OK for a firmware update but not for programming the radio beats me.

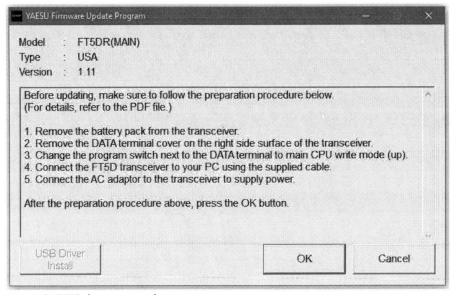

Figure 3: FT5D firmware update

1. Remove the battery from the radio

2. Lift the rubber strip covering the connectors. It does not have to be removed

3. There is a tiny switch beside the mini USB connector. Slide it up. This is a delicate procedure; I use a hat a pin and a magnifying glass.

4. Connect the USB cable between the radio and a USB port on your PC

5. Connect the plug-pack power supply. Do not reinstall the battery or attempt to turn the radio on.

6. Click **OK** on the firmware update program

If the software can see the radio on the USB port, you should see a popup window.

Click **OK** and the firmware transfer should begin without you having to click anything on the main window.

When the data transfer has been completed, new instructions appear. You need to reverse the steps you took.

1. Disconnect the plug-pack power supply

2. Disconnect the USB cable

3. Move the programming switch to the middle (normal) position. The lower position is used for a DSP upgrade, but there have been none issued so far.

4. Reseat the rubber cover

5. Connect the plug-pack power supply do not reinstall the battery

6. Do a full reset. This may not be necessary, but Yaesu says to do it. Turn the radio off. Hold down the **F MENU**, **A/B**, and **BAND** buttons while turning the radio on. A confirmation request will appear. Touch **OK** to reset the radio.

7. Enter any letter and **ENT**. Your callsign will reset when you reload your saved settings.

8. Restore your saved settings **Hold F MENU > SD CARD > BACKUP >** use the **Dial** knob to select **Read from SD > F MENU > OK > OK**. When it has been completed, press **BACK > BACK > BACK**.

9. Remove the plug-pack power supply, replace the battery, and check the firmware revision. **Hold F MENU > DISPLAY > SOFTWARE VERSION**.

FT-70D firmware update

The first step is to find out if a firmware update is required. There is no point in performing a potentially risky firmware update unless you have a major problem with the radio, or there is a bug fix update available. To check the current firmware, Hold F > select item **59 VER INF > F**. Use the **dial** to change from the CPU firmware version C 1.11 (or newer) and the DSP version D 6.04 (or newer).

Bear in mind that this radio does not have an SD card. All your settings and saved memory channels will be lost. If possible, save your configuration using the ADMS programming software before proceeding.

Navigate to the Yaesu website, find the FT-70D and click on the link. Select the files tab. Both downloads work with the DR and DE models, but there is a different firmware file for the USA version of the radio, compared to the EXP Export version. If in doubt, check the box the radio came in. My box is labelled **DST: USA** indicating a US version, so I downloaded the USA version of the firmware.

Don't bother downloading the firmware update information. It is included in the firmware update zip file.

FT-70DR/DE Firmware Update MAIN (EXP: Ver. 1.11)

FT-70DR/DE Firmware Update MAIN (USA: Ver. 1.11)

Unzip the file to a temporary directory on your PC. The 'downloads' directory is fine. The USB driver installation and instructions are in the program. Do not run the HMSEUSBDRIVER file.

Run the **FT-70D_ver111(xxx).exe** file. If this is the first time you have run the firmware update you will be asked to update the Microsoft Net framework. Even if you already have it, it is probably better just to go with the 'Download and install this feature' option. If this is the first time you have run the firmware update, click the **USB Driver Install** button. The COM port and Baud Rate boxes are not used. They can be ignored. After the driver has been installed, the **Update** button should be available. If it is not, try running the HMSEUSBDRIVER file directly.

When you click **Update** you should be presented with a screen like the one on the next page. The instructions on it are nearly right.

1. Remove the battery from the radio

2. Lift the rubber strip covering the connectors. It does not have to be removed

3. There is a tiny switch beside the mini USB connector. Slide it up. This is a delicate procedure; I use a hat a pin and a magnifying glass.

4. Connect the USB cable between the radio and a USB port on your PC

5. Connect the plug-pack power supply do not reinstall the battery

6. Click **OK**

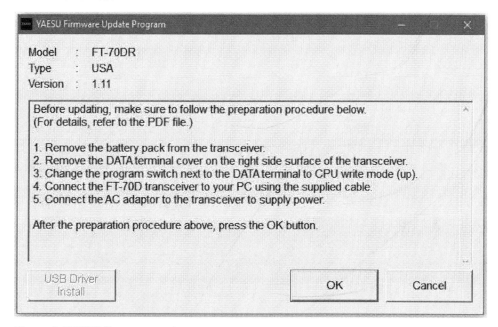

Figure 4: FT-70D firmware update

You should see a 'Select USB device,' the same as the one shown for the FT5D above. Click OK and after a slight pause, the firmware should load.

When the data transfer has been completed, new instructions appear. You need to reverse the steps you took.

1. Disconnect the plug-pack power supply

2. Disconnect the USB cable

3. Move the programming switch to the middle (normal) position. The lower position is used for a DSP upgrade, but none have been issued so far.

4. Reseat the rubber cover

5. Connect the plug-pack power supply do not reinstall the battery

6. Do a full reset. This may not be necessary, but Yaesu says to do it. You will lose all your saved memory channels. Turn the radio off. Hold down the **Mode, HM/RV, and AMS buttons and turn on the power.**

7. Enter your callsign when prompted.

8. Remove the plug-pack power supply, replace the battery, and check the firmware revision. Hold **F** > select item **59 VER INF** > **F.**

FTM-400XD firmware update

The firmware update on the FTM-400XD is similar to the other two radios, except you use the supplied SCU-20 data cable instead of a standard USB cable.

The first step is to find out if a firmware update is required. There is no point in performing a potentially risky firmware update unless you have a major problem with the radio, or there is a bug fix update available. To check the current firmware, Hold DISP > RESET/CLONE. The 'Main' firmware revision number is displayed on the top line of the screen. The current firmware is version 4.5. The DSP firmware revision is in a completely different place! Hold DISP > TX/RX > Digital. Use the Dial to scroll down to option 5 DSP version.

Note that the firmware for the FTM-400XD is not compatible with the FTM-400D.

Always backup your settings and memory channels before doing a firmware update. Usually, the radio will be reset, and all of your additions will be erased.

Hold DISP > SD CARD > BACKUP > Write to SD > ALL > OK > OK. When it has been completed, press BACK > BACK.

Navigate to the Yaesu website, find the FTM-400D and click on the link. Select the files tab. There are separate files for the FTM-400D and the FTM-400XD. Make sure you are looking at the right group. There are USA, EXP (export), and AUS versions. There is also a version for Japanese market radios, on the Yaesu Japan website. Choose the version that matches where you bought your radio. Your radio is a USA version if it has an FCC ID number on the serial number sticker. The box your radio shipped in has a DST (destination) code.

> DST: USA covers North and South America, Australia, & New Zealand

> DST: AUS covers Australia & New Zealand (check the box)

> DST: EXP/EU/CHN covers all other countries (check the box)

> The DSP firmware covers all countries. It is included in the zip file.

You only need to download the correct firmware update file. The file includes both the Main and DSP firmware updates, the update instruction manual, the Prolific USB driver and the Wires-X connection cable kit manual. Unzip the downloaded file to a convenient folder.

If you have never used the SCU-20 data cable, you will have to run the prolific USB driver installation. It will have to be unzipped first. The file you need to run is called PL23XX-M_LogoDriver_Setup_402_20220309.exe. Launch Windows Device Manager from the Windows search box, and look in the Ports (COM & LPT) section for a COM port labelled 'Prolific USB-to-Serial Comm Port.' Take note of the com port number.

OK, once the software can see the radio connected to your PC over the SCU-20 cable, we can FINALLY get to the job at hand!

To update the Main firmware

- Did you make a backup onto the SD card?

- Connect the radio to the PC via the SCU-20 data cable

- Run the FTM-400XD_ver450(xxx).exe file

- Set the COM port dropdown box to the Prolific comport number

- Click **Update**

- Disconnect the power from the radio by unplugging the red and black power cable. Yes, you do have to do this step.

- There is a small rubber bung on the speaker side of the radio, near the mic connector. Remove it. Change the switch under the cover to **Update** by gently sliding it towards the back (power connector) end of the radio. Use a very small screwdriver or a similar pointy object

- Plug the DC supply back in. Do not turn the radio on or push any buttons. The screen should be black

- Press **OK** on the update software. After a pause, the firmware should start loading. It is slower than updating firmware on handheld radios, taking about two minutes to complete. This will be because data transfer over a data cable is much slower than a direct USB connection.

- When it completes, reverse the procedure as described in the software window. Disconnect the power, disconnect the SCU-20 cable, reset the switch, replace the rubber bung, and plug in the power.

- Click **Exit** in the software.

- Turn the radio on and perform a 'Factory Reset.' **Hold DISP > RESET/CLONE > FACTORY RESET > OK?**

- Reload your saved settings. **Hold DISP > SD CARD > BACKUP > Read from SD > ALL > OK > OK.** When it has been completed, press **BACK > BACK > BACK.**

- Recheck the firmware version number in the Reset/Clone menu.

To update the DSP firmware

The zip file contains a 'PCTOOLe_v0431' folder. Or a newer number. Run the PCTOOLe_vxxxx.exe file.

- Did you make a backup onto the SD card? This process should not affect your settings, but better safe than sorry.

- Turn the radio **off**

- Connect the radio to the PC via the SCU-20 data cable

- Set the COM port dropdown box to the Prolific comport number

- Click **OK**

- The instructions say that the cable must be plugged in, and the power cable must be connected to the radio. Then it says to turn the radio on in 'C4FM-DSP F/W write mode,' but it does not say how to do that!

- Hold down the **DX** and **F MW** keys while turning on the radio with the **power** button. The radio should start with a 'DSP UPDATE' message followed by 'PUSH D KEY.'

- Push the **DX** key and then click **Update** on the software.

- When the DSP firmware has loaded the radio will display 'WRITE END.' Turn the radio off. Remove the SCU-20 cable unless you are using it for the PDN Wires-X mode. It says to reconnect the power cable, but of course, it is already plugged in. **Exit** the software.

- Turn the radio back on. It should start normally. You can check the new DSP firmware number in the **Digital > DSP version** menu.

YAESU SERIAL NUMBER IDENTIFICATION

It may be of interest to identify the date range of your Yaesu transceiver. Since we don't know how many radios are made in each lot or even if that is a constant, we cannot calculate how many radios have been made.

Digit	Function	Range
1	Last digit of the year of manufacture	1 = 2021, 2 = 2022, 3 = 2023, 4 = 2024 etc.
2	Month of manufacture	C=Jan, D=Feb, E=Mar, F=Apr, G=May, H=Jun, I=Jul, J=Aug, K=Sep, L=Oct, M=Nov, N=Dec
3,4	Lot number	01 to 99
5,6,7,8	Serial number	0001 to 9999

C4FM technical information

Yaesu digital voice radios use C4FM modulation. The emission designator is 9K36F7W, indicating that C4FM requires a transmission bandwidth of 9.36 kHz inside the 12.5 kHz RF channel. There are three digital sub-modes, and of course, the radios can also receive and transmit analogue FM.

The normal DN (digital narrow), 'voice plus data' (V/D) mode, transmits digital voice data, synch and callsign data, and FEC (forward error correction).

The VW (voice wide), 'voice full-rate (VFR) mode, transmits voice data over the full bandwidth. This results in higher fidelity (better quality) speech, but there is no forward error correction. It will provide very good audio provided there is a strong signal. Generally, this mode will be excellent across a local repeater. The DN mode is better over internet-linked calls.

The DW (digital wide), 'data full rate' (DFR) mode, is used for sending data such as pictures. It takes the full bandwidth to send data at a higher rate than the DN mode, but you cannot use the radio for voice at the same time.

Analogue FM can be used if the signal is too weak for digital signals. The radio uses significantly less power on FM so switching to FM can help if you are getting low on battery reserve. There is a normal FM mode and a narrow FM mode.

MODULATION

Yaesu uses the terms C4FM and FDMA to describe the modulation used for YSF. C4FM stands for continuous four-state frequency modulation. It was originally designed for P25 (phase 1) commercial radio systems. But it is not used for phase 2 and phase 3 P25 systems. C4FM is a variant of 4-state FSK (frequency shift keying) It is referred to as 'continuous' because the radio transmits continuously while the PTT is pressed, rather than the bursts of data that a DMR radio transmits. This stems from the use of FDMA for YSF and TDMA for DMR.

FDMA stands for frequency division multiple access. Yaesu Fusion repeaters only transmit one digital voice channel, so I assume that in this case, FDMA refers to the way the data and digital voice signals are interleaved in the voice plus data (V/D) mode. The radio transmits seventy-two bits of voice data followed by seventy-two bits of data and FEC. This interleaving does not apply to the VFR and DFR modes which don't have any FEC. TDMA stands for time division multiple access, meaning that DMR transmits a voice channel on 'time slot 1' followed by a second voice channel on 'time slot 2.'

C4FM transmits on four frequencies centred around the nominal channel frequency. Every two bits of the digital data stream creates a signal on one of the four frequencies.

According to 'Amateur Radio Digital Standards,' published by Yaesu in 2013 and revised in July 2015, the offset frequencies are different to P25. Yaesu C4FM transmits the displayed frequency +900 Hz, +2700 Hz, -900 Hz, and -2700 Hz.

The data sent to the modulator is 'differential' or 'Gray' coded in such a way that a long string of zeros or ones does not result in the same frequency being transmitted for an extended time. Each 0.2 ms transmission of one of the four possible frequencies is called a symbol, and each symbol carries two bits of information. This means the bit rate of the system is two times the symbol rate. The symbol rate is 4800 bauds, and the bit rate is 9600 bits per second. In the DN mode, 4400 bps is voice data generated by the AMBE voice codec, 2800 bps is forward error correction, and 2400 bps is used for signalling, synchronisation, and control functions. *TIP: 4800 bauds means 4800 symbols per second.*

The data is sent in 960-bit sub-frames. A 'frame' consists of a 960-bit header, up to seven 960-bit communications sub-frames, and a 960-bit 'terminator.' Each 960-bit sub-frame takes 100 ms to send because the overall data rate is 9600 bps.

Each of the communications sub-frames has 20 bits of frame sync and 200 bits of FICH 'frame information channel' data, followed by five 72-bit data words interleaved with five 72-bit voice data words. The FICH data tells the network what information is included in the data sub-frames.

VOICE CODING

When you transmit, your voice is converted into a digital data stream of binary bits which is applied to the C4FM modulator in the transmitter. The conversion is done with a propriety chip running the DVSI (Digital Voice Systems Incorporated) AMBE+2 Vocoder. A Vocoder is a CODEC (coder-decoder) designed for voice signal digital coding, compression, multiplexing, and encryption.

In the receiver, the C4FM signal is demodulated back to a data stream and the AMBE+2 Vocoder converts it back to an audio signal, ready for the audio amplifier and speaker, Bluetooth link, or headphones.

The voice coding has nothing to do with the C4FM modulation. The only requirement is that the modulation must be able to keep up with the voice coding so that no voice data is lost.

CLASS C AMPLIFIERS

A C4FM-modulated radio only transmits one frequency at a time. Each transmission is at full power. The radio also transmits analog FM at full power, regardless of the deviation. Because there is no amplitude modulation in either mode you do not need a linear power amplifier. If you want to boost the signal, you can use a more efficient class C amplifier rather than a traditional class AB linear amplifier. Of course, you can use a linear amplifier if you have one.

THE DIFFERENCE BETWEEN FSK AND PSK

With FSK (frequency shift keying) the binary data signal is represented by changes in the RF frequency. In PSK (phase-shift keying) the frequency remains the same and the binary data is represented by changes in the phase of the signal. Each phase or frequency state is known as a symbol. The symbol rate for 2-state frequency shift keying 'FSK' or 2-state phase-shift keying 'BPSK' is the same as the bit rate because there are only two possible frequency or phase states, each representing one binary bit. Note that it is not the current frequency or phase that determines whether the symbol represents a binary one or a zero. It is the change of state.

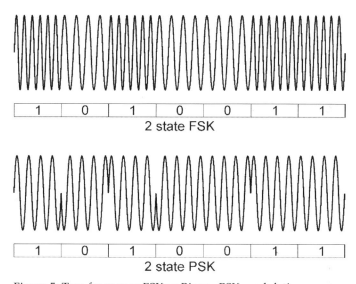

Figure 5: Two frequency FSK vs Binary PSK modulation

There are scrambling and data coding techniques aimed at eliminating long runs of ones or zeros. Many systems use differential coding where a change of state means that the current binary bit is different to the previous bit. No change of state indicates that the current binary bit is the same as the previous bit.

4FSK (four-state frequency shift keying) uses four frequencies rather than two frequencies. 4PSK (four-state phase shift keying), also known as QPSK, 'quadrature phase-shift keying,' uses four phase changes, 0°, 90°, 180°, and 270°, rather than the 0° and 180° phase shifts used in BPSK (binary phase-shift keying).

When you use 4FSK or QPSK modulation, each phase or frequency state 'symbol' carries 2 bits of the digital signal, so the bit rate is twice the symbol rate. Using this higher order of modulation allows you to transmit twice as much data in the same amount of time. But it requires more bandwidth to do so. The data is usually manipulated so that all four states are transmitted regularly. This is particularly important for QPSK because it affects the shape of the transmitted spectrum and clock recovery at the receiving station.

COMPARISON OF DIGITAL VOICE MODES

All of the digital voice systems in common use on the VHF and UHF amateur radio bands use digital forms of FM modulation. This is because it has been traditional to use FM at those frequencies and because commercial operators are interested in migrating customers from FM repeaters onto digital repeaters. It is good because it means that all YSF radios support FM operation and can access FM repeaters.

The C4FM modulation used for YSF and the 4FSK modulation used for DMR produce a four-state constellation and the bit error rate (BER) performance is nearly identical. Both modes are about 1 dB worse than QPSK at the same rate. The GMSK modulation used for D-Star is slightly better. YSF and DMR systems require a received signal two or three dB stronger to achieve the same BER as D-Star.

All of the digital voice modes in this comparison transmit extra bits to perform FEC (forward error correction) which improves the voice quality by repairing errors in the received data.

RF factors	DMR	D-Star	Fusion	P25 phase1
Vocoder	AMBE+2	AMBE+2	AMBE+2	IMBE
FEC	Yes	Yes	Yes	Yes
Modulation	4FSK	GMSK	C4FM	C4FM or CQPSK
Emission	7K60FXE	6K00F7W	9K36F7W	8K10F1E
Transmission rate per voice channel	4800 bps	4800 bps	9600 bps	9600 bps
Mod Bandwidth	7.6 kHz	6.0 kHz	9.36 kHz	8.1 kHz
Channel bandwidth	12.5 kHz	6.25 kHz	12.5 kHz	12.5 kHz
Voice channels	2	1	1	1
Developer	ETSI	JARL	Yaesu	APCO

User information	DMR	D-Star	Fusion
Registration required	Yes	Yes	No [*1]
User ID transmitted	DMR ID[*2]	Callsign	Callsign
User ID displayed	DMR ID[*3]	Callsign	Callsign
Is the ID OK for FCC and other jurisdictions?	No	Yes	Yes

*1 You do not have to register to use a Wires-X through a Fusion repeater, or for access to the public YSF, FCS, and YCS reflector systems. You do have to register if you want to use the Wires-X software in PDN/HRI mode or operate an HRI-200 interface on the Wires-X network.

*2 most amateur radio licencing authorities require you to transmit your callsign. DMR does not do this automatically. You need to identify your transmission by saying your callsign, the same as you would on an FM repeater.

*3 a mobile, handheld, or hotspot will only display the other station's callsign if it is held in the contact list in your radio, or the Talker Alias includes the callsign. Otherwise only the ID number is displayed. There are around 228,699 registered DMR users. Few radios can store all of them. With other radios, you will have to be more selective. You can download selected data from https://www.radioid.net/. Or you may elect to only enter contact data from your local area, or for amateur radio operators you call often.

User repeater connections	DMR	D-Star	Fusion
Talk to local stations	Yes	Yes	Yes
Link to another repeater	No	Yes	No
Internet-linked systems	Talk groups	Reflectors	Wires-X rooms Reflectors
Selection	Channel switch	UR field	Room name or reflector number
Private Call	Yes	Not usually	No
Echo / Parrot test	Yes	Yes	Yes

Ease of use	DMR	D-Star	Fusion
Memory selection	Channel switch	Dial or list	Dial or list
Analogue mode	Programmed channel	Button	Button or AMS
Programming	Difficult*5	Medium	Medium *4
Using the radio	Easy	Medium	Medium
Radio models (mostly)	Many	Icom	Yaesu
Radio cost	Cheapest	Medium	Medium
Hot spots	Any MMDVM Pref Duplex	Any MMDVM	Any MMDVM or Wires-X
Send photos	No	Yes	Yes
SMS (text) messages	Yes	Yes	Yes (Wires-X)
Send GPS/APRS data	Some models	Yes	Yes

*4 programming channels and features using the controls on the radio is fairly easy, if not always intuitive. I found getting the ADMS programming software to work very challenging.

*5 It's not that programming a DMR radio is difficult to accomplish. You just download a new code plug config file using a USB cable.

The annoying thing that makes life a little difficult is that you have to edit and download a new file any time you want to add a new talk group or repeater to the radio. Adding a new repeater can involve adding many additional lines to the file. It is not as easy as travelling to another city and entering the local repeater channel frequency and CTCSS tone from the radio keypad the way you would for an FM repeater. Most DMR radios can be programmed with the radio keypad, but it is fiddly.

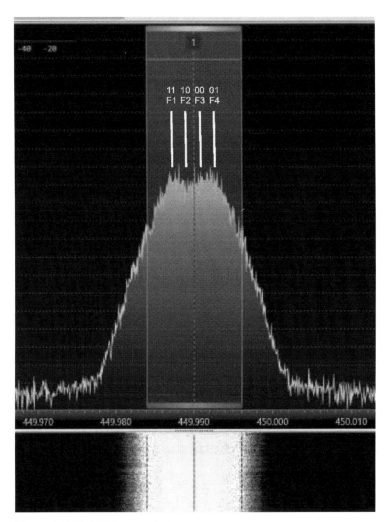

Figure 6: YSF C4FM transmission

This spectrum plot of a C4FM transmission was created by Rob Robinette K9OJ. I have added the four C4FM frequencies so that you can see how the RF spectrum is generated. The binary numbers don't represent the original data stream because of the differential coding which ensures that each frequency is transmitted regularly.

https://robrobinette.com/Ham_VHF-UHF_Digital_Modes.htm

Getting started

The first thing you need to do is check out what Yaesu Fusion or non-Yaesu digital voice repeaters are available in your area. It is a good idea to talk to YSF users at your local amateur radio club. If there are no Fusion or compatible digital voice repeaters in your area, but you have good a good internet connection, you can buy or build a hotspot and use that to communicate with radio hams on the YSF reflectors. Alternatively, you can connect some Yaesu C4FM radios directly to your home network and operate in the PDN or HRI 'terminal' modes.

Next, you need a Yaesu digital handheld, mobile, or desktop radio. Take a look at some of the things to look out for, on page 6. I have concentrated on the FT-70D, FT5D, and FTM-400XD models. The Yaesu FTM-400XD is the same as the original FTM-400D, with an improved GPS receiver. Programming and using other Yaesu radios should be similar to at least one of those models. The book includes setup instructions, and how to connect to rooms and reflectors. Once you have purchased a radio, there is very little work to do before you can use it. You need to enter and store the frequencies of your local repeater or hotspot, as you would for an FM transceiver, and you need to enter your callsign. After that, you are "good to go." It may seem rather daunting at first because of all the new terms you need to learn and having to adjust to a different way of operating the radio. But we will step through the process, and it will soon become second nature.

Unlike DMR where every talk group must be programmed for every RF channel, YSF radios can access all YSF reflectors and Wires-X rooms without you having to download any database files. A YSF radio only needs one channel for each repeater or hotspot. You can travel to a different town or country, enter the frequencies of the local Fusion or non-Yaesu repeater and use your favourite rooms or reflectors very easily. You don't have to download a callsign database either. The radio will display the callsign and name of anyone you are hearing.

I have described changing settings directly on the radio and using the free Yaesu ADMS configuration software. Some functions are easy to change using the radio buttons, others, like adding a lot of repeater channels are easier using the software.

ENTER YOUR CALLSIGN

The very first thing that you must do is enter your callsign into the radio. No digital mode conversations can be made unless this simple step is done.

FT5D F Menu > Callsign > enter callsign > Back > Back

FTM-400XD Disp Setup > Callsign > Change > enter callsign > ENT> Back > Back

FT-70D F > 64 My Call > enter callsign > F

Using a YSF reflector

I am assuming that you have programmed a channel for your hotspot or repeater, and you have your radio set to that channel now.

You cannot access YSF reflectors via a Fusion repeater or in the PDN/HRI Wires-X terminal mode.

Press the DX, GM/X, or F > AMS keys to connect to the node.

TIP: You cannot unlink from, or link to, a reflector if someone is talking on it. Wait 20 seconds and try again. If you are having difficulty linking to a reflector, try unlinking first. If that fails, the reflector might be offline.

UNLINKING

Sometimes it is necessary to unlink a reflector or Wires-X room before connecting to another one. If you are using a repeater, it is polite to leave it unlinked if you previously established a link and you now wish to inform others that you have finished with it. Leave the link in place, if the repeater was already linked to the room or reflector, you have been using.

On the FTM-400XD and similar radios, press and hold the asterisk key * on the microphone until the radio transmits. You will see a reply come back from the repeater or hotspot. On the handheld radios such as the FT5D and FT-70D, press and hold the BAND button until the radio transmits.

LINKING TO A YSF REFLECTOR

FT5D

Figure 7: FT5D connected screen

This image shows the radio connected to YSF node 41562 '001-CQ-UK.'

The middle line shows the callsign of the hotspot or non-Yaesu reflector. The line at the bottom shows the connected YSF reflector. Flashing text indicates the last used reflector and that it is not currently linked. Touch the reflector name to re-link to it.

Touch **Search & Direct** to link to a different reflector.

Use the Dial or touch one of the five category banks to select a previously saved reflector.

Touch ALL to select from all the available reflectors. Note that these are loaded in batches.

Or touch Search & Direct again to do a text or number-based search.

A context-based text search is usually successful. Sometimes it just comes back with "no data." Some searches that worked for me are AMERICA, UK, YSF, XLX, STAR, VK, and NZ.

An alternative method is to use the ID search. But you need to know the reflector ID number. Touch Search & Direct then ID in the top left corner of the alpha keypad. Enter the DTMF ID number of the YSF reflector after the # symbol. For example, 41562 for the '001-CQ-UK' reflector. The DTMF ID numbers are available at https://www.pistar.uk/ysf_reflectors.php

FTM-400XD

The process for the FTM-400XD is the same although the screens look a little different. You can only use the A-band (upper display) for digital voice.

Figure 8: FTM-400XD connected screen

This image shows the radio connected to YSF node 41562 '001-CQ-UK.'

The middle line shows the callsign of the hotspot or non-Yaesu reflector. The line at the bottom shows the connected YSF reflector. Flashing text indicates the last used reflector and that it is not currently linked. Touch the reflector name to re-link to it.

The red number is the number of nodes (people & repeaters) connected to the reflector.

Touch the **white down arrow** to link to a different reflector.

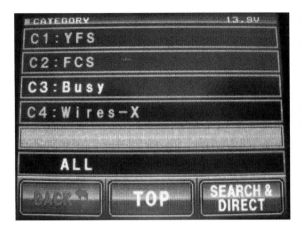

Use the **Dial** or touch one of the five **category banks** to select a previously saved reflector.

Or touch **ALL** to select from all the available reflectors. Note that these are loaded in batches.

Touch **Search & Direct** to do a text or number-based search.

The context-based text search is very often unsuccessful. Often it just comes back with "no data." The only search that worked for me was AMERICA.

*TIP: annoyingly you can't back out of this screen without erasing the search text first. Then the **Back** key appears, and you can exit the screen!*

FTM-400XD linking using a DTMF ID number

The radio does not have an ID search function. The 123 soft key is labelled with the required # symbol, but it is missing from the onscreen keyboard. However, all is not lost. You can enter a DTMF ID from the normal 'connected' screen. Press and hold the **#** key on the microphone. The display will change to DTID #. Enter the **DTMF ID** number using the keys on the microphone then press the **#** key again. The radio will link to the selected reflector. The DTMF ID numbers are available at https://www.pistar.uk/ysf_reflectors.php

TIP: this mode is difficult to get out of if the connection fails. The easiest way is to enter a different DTMF ID number then #. The only other way seems to be to power off the radio.

FT-70D

The FT-70D does not have a search function. As usual, use the **F > AMS** keys to connect to the node. If you want to select a different reflector turn the **Dial** to show En *. Enter the YSF DTMF ID (not the YSF number), from the list at https://register.ysfreflector.de/ or https://www.pistar.uk/ysf_reflectors.php. The list includes linked XLX reflectors as well as YSF reflectors.

Pi-Star Hotspot

You can link a Pi-Star hotspot to a YSF reflector from the admin tab of the Pi-Star dashboard. Select the required reflector using the dropdown box in the 'YSF Link Manager' section and click **Request Change**. There is a live search box at the top of the list. It updates the list as you type in letters or numbers. Try XLX or America.

Using an FCS reflector

The FCS reflectors are a little different to the YSF reflectors. The system consists of 24 servers each identified with a three-number code. A server can have up to 100 modules, numbered 00 to 99. Each FCS reflector has a dashboard page on the internet. You can find links to them at http://xreflector.net/. You can link your radio or hotspot using a combination of the server number and the module number. For example, linking to FCS53089 connects you to FCS530 module 89. If you go to the server dashboard and click the DG-ID tab, you will see a list of the modules that the server supports. Modules 1 to 7 are common to most FCS servers, but the other numbers may link to completely different groups. Clicking on the **Repeater** tab shows you the connected callsigns and the module they are using. If a callsign has several modules listed, it indicates they are listening to static modules. This is a great way to see what activity there is on the various connected networks.

You cannot access FCS reflectors via a Fusion repeater or in the PDN/HRI Wires-X terminal mode, and the same rules apply regarding connections while the reflector is in use. Press the **DX**, **GM/X**, or **F > AMS** keys to connect to the node.

FT-70D

If you are using a Pi-Star hotspot, there is no "official" way to access FCS reflectors from the FT-70D. That said, some of them are accessible. The problem is that the DTMF codes do not match the reflector numbers, and some of them clash with the numbers used for the YSF reflectors. I do not have the time to find codes for more than 1000 active FCS modules, but I have worked out a few. (Table left side).

If you are using the EA7EE version of Pi-Star on your hotspot, the hotspot interprets the code and selects the requested FCS reflector. (Table right side).

Turn the **Dial** to show En *. Enter the number from the table below. Then press **AMS**.

Pi-Star Code	Reflector	EA7EE Code	Reflector
00010	FCS00100 Repeater	53002	FSC53002 Europe
00011	FCS00101 Deutschland	53021	FSC53021 Worldwide
00012	YSF 00012 Japan Link	53035	FSC53035 UK-1
00013	FCS00102 Worldwide	53037	FSC53037 America Link
00014	FCS00103 Switzerland	00200	FSC00200 Talk-USA 1
00015	FCS00104 Denmark	00201	FSC00201 Talk-USA 2
00016	None	00301	FSC00301 Canada French
00017	YSF00017 XLX017	20810	FSC20810 WW German
00018	FCS00106 USA	20821	FSC20821 Worldwide

FT5D

You can use the Search & Direct function to connect to FCS reflectors. It is often best to unlink first.

Connect to the node using the **GM/X** key as usual. Touch **Search & Direct** > **Search & Direct**. Enter **FCS** and the start of the reflector number. Entering **FCS** will show you FCS00100 to FCS00119. Entering **FCS202** will show you FCS20200 to FCS20219.

Pressing the **white down arrow** will load the next 20 reflector numbers.

If the reflector you choose, links successfully, you should get the information page. You can touch **ADD** to add the reflector to one of the five category banks, or the **BACK** key to exit.

To disconnect a reflector, connect to the node using the **GM/X** if you are not already connected. Hold the **Band** button down until the radio transmits. You should get a Morse code H, a 'Not Connected' popup message, and the reflector display will flash.

FTM-400XD

You can use the context-based search function to list and select FCS reflectors. In the connected mode touch the **white down arrow** > **Search & Direct**. Enter **FCS** to jump to the top of the FCS list, or **FCS5** to get to the FCS500 reflector etc. Make a selection to connect to the reflector. Unfortunately, the ADD function does not work with reflectors on the FTM-400XD, so press **BACK** to exit.

TIP: sometimes the search function works and other times it does not. I have no idea why.

TIP: I find the beep when you press any key to be obnoxiously loud. I don't need a beep to tell me that I pressed a key! Unfortunately turning off the beep also turns off the chimes that inform you that the reflector connected successfully.

You have to be in the connected mode to disconnect a reflector. Press and hold the asterisk key * on the microphone until the radio transmits. You will see a 'Not Connected' indication on the radio. The reflector changes to a slightly grey colour.

Pi-Star Hotspot

You can link a Pi-Star hotspot to an FSC reflector from the admin tab of the Pi-Star dashboard. Select the required reflector using the dropdown box in the 'YSF Link Manager' section and click **Request Change**. The FCS reflectors are listed after the YSF and XLX reflectors, so it is easier if you shorten the list by typing FCS into the search box at the top of the list. The search function updates the list as you type in letters or numbers. Enter **FCS** to jump to the top of the FCS list, or **FCS3** to get to the FCS300 reflector, and so on.

Using a YCS reflector

The YCS reflector system is new and not fully supported by the 'standard' version of Pi-Star. Its primary function is to link Yaesu repeaters, Yaesu YSF and FCS reflectors, and other digital modes. YCS reflectors are always connected to the DMR+ and Brandmeister DMR networks and many XLX reflectors.

The cool thing about the YCS servers is that they are designed to work like DMR. You can configure your hotspot (or BlueDV) to listen for several modules on the YCS server at the same time. For example, you could listen for calls on your local reflector, CQ-UK, and America Link. You will hear a conversation on an active module. The audio will not transfer to a second module until the one you are currently hearing has been idle for 20 seconds. This is to stop you from losing the link mid-conversation.

There is a 'fork' of the Pi-Star software created by Manuel Sanchez Raya EA7EE http://pistar.c4fm.es/ which supports the YCS 'DG-ID' mode. Download at https://drive.google.com/file/d/1jU6Bia-DDkdH4bePjGDn_Iph6qMdOrts/view.

BlueDV also supports the 'DG-ID' mode. You have to add an options line to the .ini configuration file. It is quite easy. See 'BlueDV setup for YCS reflectors' on page 67. Option numbers less than 7 link to the same reflectors on almost all of the 20 YCS servers. Option numbers greater than 7 may link to different reflectors if you change to a different YCS server.

The **DV Matrix** links the 20 YCS servers together. Have a look at the Server tab on its dashboard. It shows the modules in use on each of the servers.

YCS-Network System Fusion II (YSF/FCS/IMRS)
YCS Information
DVMatrix
YCS202 (GR)
YCS207 (NL)
YCS208 (FR)
YCS222 (IT)
YCS224 (EA)
YCS226 (RO)
YCS232 (AT)
YCS235 (GB)
YCS259 (MD)
YCS260 (PL)
YCS262 (DE)
YCS268 (PT)
YCS284 (BG)
YCS310 (US)
YCS311 (US)
YCS334 (MX)
YCS450 (KR)
YCS505 (AU)
YCS530 (NZ)
YCS724 (BR)

Accessing a YCS server

You access the YCS network by linking to an FCS reflector and module number that is hosted on a YCS server. For example, linking to FCS50550 links to module 50 'Australia' on YCS505 'C4FM Australia.' You should see your callsign listed on the reflector dashboard. If you are using the EA7EE version of Pi-Star or BlueDV with options added, you will hear all the modules that you have selected. Otherwise, you will only hear the module that you linked to. In this case module 50.

The process is exactly the same as linking to any other FCS reflector. The image on the right shows the dashboard links at http://xreflector.net/.

Sometimes you cannot link to a module if another module is active.

FT5D

Entering YCS into the search function on my FT5D reveals four YSF reflectors that have YCS in their name. They are linked to the YCS network so if you link to them you should appear on the relevant YCS dashboard.

00-YCS235 Multi Networks Two	on YSF reflector YSF 41619
GB-CHAT2-Y TG2352 CQ UK DMR	on YSF reflector YSF 23502
GB-CQ-UK-Y YCS235	on YSF reflector YSF 23501
NL-NL-YCS2 YCS207	on YSF reflector YSF 20701

Otherwise, just enter the FCS server and module number. For example, a search for FCS235 will list the first 20 modules. A search for FCS23535 lists 'CQ-UK.'

FTM-400XD

A search for 'YCS' fails to find any reflectors. You can search for the FCS reflectors on the YCS servers. For example, a search for FCS235 will list the first 20 modules.

Pi-Star hotspot (EA7EE)

A standard Pi-Star can link to a single YCS server by linking to the FCS reflector and module. The EA7EE version of Pi-Star supports the YCS feature for listening to multiple static modules.

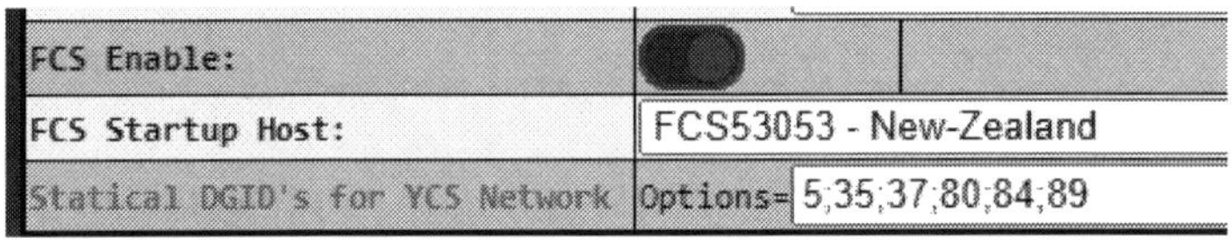

Figure 9: EA7EE static module options

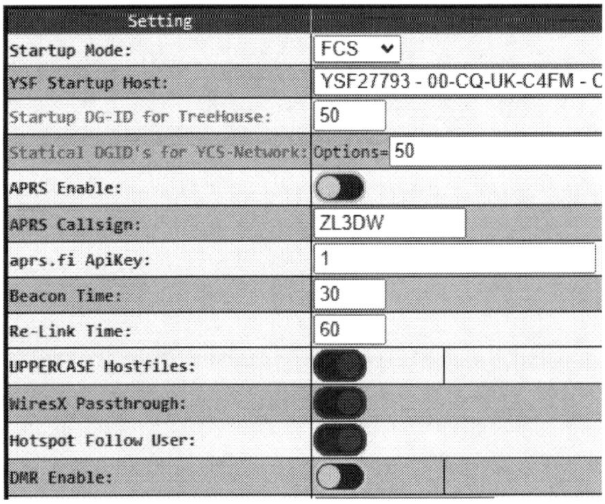

The setup screen is slightly different. You have the choice of the hotspot starting up linked to a YSF reflector, or an FCS reflector, which can be a YCS one. I elected to start connected to module 53 'New Zealand' on FCS530. I have set 'options' for six static modules: Oceania, CQ-UK, America Link, NZ Canterbury repeater, ZL Chat, and the ZL IPSC2 DMR server.

Figure 10: EA7EE Fusion setup

Using Wires-X

Using Wires-X is much the same as connecting to the reflectors through a repeater or hotspot. But in this case, the reflectors are 'rooms' hosted on the Yaesu network. You can connect to Wires-X rooms through a repeater without loading the Wires-X software or registering with Yaesu. You do have to register to use the software for the PDN or HRI terminal modes or operating through a radio connected to an HRI-200 interface. The dashboard is the Wires-X software that runs on your PC.

The Wires-X software is covered in the next chapter. The HRI-200 in the chapter after that, and the PDN mode in the chapter after that.

WIRES-X THROUGH A FUSION REPEATER

Accessing Wires-X rooms is very easy if you are lucky enough to have a Yaesu Fusion repeater in range. I can trigger the Fusion repeater at My Grey. You can access any Wires-X room without having to load a large database onto your radio, and you only need one channel in the radio memory bank, for the repeater.

The main disadvantage of using a repeater rather than a terminal mode connection is that you have to be mindful of other repeater users. Especially when changing the room that the repeater is linked to. The main advantage of using a repeater is that it is easy to talk to local hams in your area. And you can use the repeater while you are mobile or portable.

TIP: it is not possible to access the Wires-X network through a hotspot or DV dongle. You can access some Wires-X rooms which happen to be connected to YSF, FCS, YCS, or XLX reflectors, but there is no direct access. You must use a Yaesu radio. With some networking options, you need two Yaesu radios.

WIRES-X THROUGH A CONNECTED NODE

If you don't have a local Fusion repeater, or you want more control over how you access the Wires-X rooms, you can opt for terminal mode operation. I think that nearly everyone will choose to use the PDN method because buying an HRI-200 costs more money and does not add very much capability. In either case, you must tether a Yaesu digital radio to a USB cable attached to your PC, and the PC must be running the Wires-X software. If you want to wander around the shack or operate mobile in your local area, you need a second Yaesu digital radio to work as an access radio. The requirement for the PC to be connected and running Wires-X, plus effectively wasting a radio for use as a hotspot is a major downside. It is very old-fashioned compared to using an MMDVM hotspot which connects directly to your WiFi network and does not need your PC. The HRI-200 should have been updated with an RF stage so that it does not need a donor radio, and it should have an Ethernet port and WiFi so that it doesn't require software running on your PC.

UNLINKING

Sometimes it is necessary to unlink a Wires-X room before connecting to another one. If you are using a repeater, it is polite to leave it unlinked if you previously established a link and you now wish to inform others that you have finished with it. Leave the link in place, if the repeater was already linked to the room or reflector, you have been using.

On the FTM-400XD and similar radios, press and hold the asterisk key * on the microphone until the radio transmits. You will see a reply come back from the repeater or node. On the handheld radios such as the FT5D and FT-70D, press and hold the **BAND** button until the radio transmits. A flashing or dimmed room indication on the connected radio display indicates the room is not linked. You can re-establish a link to the same room by touching the room name on the display.

TIP: You cannot unlink from, or link to, a room if someone is talking on it. Wait 20 seconds and try again. If you are having difficulty linking to a room, try unlinking first. If that fails, it might be offline.

USING THE SEARCH FUNCTION

FT5D

You can operate the FT5D as an access radio without going into the connected mode. Simply set the radio to the frequency and offset stated on the Transceiver settings tab in the Wires-X software and set the DN mode. You can use simplex by setting the offset to zero on the radio and in the software. Select a room in the Wires-X software, right click and choose **Connect**. If you want to use the search function on the radio, press the **GM/X** key to connect to the node.

This image shows the radio connected to America Link on Wires-X. That room is also available on YSF and FCS reflectors through bridging links.

The middle line shows the callsign of my Wires-X PDN node. The line at the bottom shows the connected Wires-X room. Flashing text indicates that the last used room is not currently linked.

Touch **Search & Direct** to link to a different reflector.

Use the **Dial** or touch one of the five **category banks** to select a previously saved room. Only Wires-X rooms will be displayed.

Touch **ALL** to select from all the available rooms. Note that these are loaded in batches.

Or touch **Search & Direct** again to do a text or ID number-based search.

The context-based text search works better on Wires-X. Some searches that worked for me are AMERICA, UK, CQ, and ZL. The radio should display the information and news page when the room connects. Touch the **ADD** icon at the bottom to add the room as a favourite to your category memory bank. This is the only opportunity you will get to do it.

An alternative way is to use the ID search. But you need to know the DTMF ID number. This is available on the Wires-X software, but if you looked it up there, you might as well connect from there. Touch **Search & Direct** then **ID** in the top left corner of the alpha keypad. Enter the **DTMF ID** of the Wires-X room after the # symbol. For example, 27793 for the '-----------CQ-UK' room. The ID search function does not allow you to add the room to a category memory bank.

FTM-400XD

If you are using the FTM-400XD as the donor radio in PDN mode and you are also using a second radio as an access radio, you cannot change settings, search, or transmit on the FTM-400XD.

Press the lower **Dial** knob to change from access radio mode to direct mode. The display will show the access frequency in access mode, and DIRECT in direct mode. In direct mode, you will hear incoming calls on the radio and be able to transmit back to them. The only other live control is the volume knob.

If you are using the FTM-400XD as the donor radio in HRI-200 mode, you must use a second radio as an access radio. On the upper screen, the FTM-400XD will display the frequency it is using to receive the access radio. On the lower part of the screen, it will display the frequency it is using to transmit to the access radio. You cannot change settings, search, or transmit on the FTM-400XD.

If you are using an FTM-400XD as an access radio, the search functions work the same as described in the 'linking to a YSF' reflector section.

Wires-X software

I found it very difficult to get the Wires-X software to work. I think that this was mostly due to configuration issues regarding the HRI-200. If you are configuring the software for the PDN/HRI mode it should be a much smoother experience. At least the program worked well once I had it configured.

The same software is used for the PDN/HRI mode or when using an HRI-200. You do not have to reinstall the software provided you have V1.54 or newer installed. It is very easy to swap between the HRI-200 mode and the PDN mode.

GETTING STARTED WITH WIRES-X

The HRI-200 comes with a CD containing a copy of the Wires-X software, but it is better to download the latest version. If you are using the PDN/HRI mode, you will have to download it anyway. You will not be able to find or install the Wires-X software without first registering with Wires-X. After that has been completed you will gain access to the 'Node owner's page' on the Wires-X website. https://www.yaesu.com/jp/en/wires-x/index.php

First complete registration on the Wires-X website. You will need either the Node ID of your radio or the serial number of your HRI-200. Then you fill in some contact details and choose an ID and password. After a day or two, you will receive an email with a link to the Wires-X node owner's page a Node ID and a Room ID.

Your radio ID is in the main menu. Press **F Menu** or **DISP/SETUP > GM > Radio ID Check**. The HRI-200 serial number is on the bottom of the unit. The FT-70D can't do PDN mode or interface with an HRI-200 but it does have a radio ID under **F > 43**.

Installing the software

When you have your logon ID and password and the 'node ID' you can install and run the Wires-X software. The node owner's page has several links. Download WIRES-X PC Software (Ver.1.540), which is the Wires-X software zip file. The zip includes 32-bit and 64-bit setup files, an HRI-200 manual, a Wires-X PDN manual, a Wires-X radio ID check document, and the Wires-X installation manual. Don't bother downloading the WIRES-X PC Software Update Information (Ver.1.540) document. The WIRES-X Remote Monitor Software (Ver.1.010) program is for remote controlling the Wires-X software from a remote computer. I have not used it, so it is outside the scope of this book.

The webpage also includes a link for doing a firmware upgrade on the HRI-200. I don't know if it is necessary. I had so much trouble getting the program to recognise that the HRI-200 was connected, that I ran the firmware upgrade out of desperation.

Below the firmware file there are links to several manuals that are included in the program zip file, so don't bother to download them again. Finally, there is a note about setting internet connections for the HRI-200 and an 'Initial Setup' note for the HRI-200. This information is in the HRI-200 manual included in the program zip file.

Configuring the software

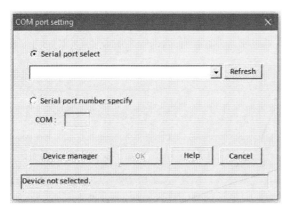

The first time you start the software you will probably get a 'COM port setting' popup box like this. Make sure that the radio data cable or the HRI-200 USB cable is connected between the unit and a USB port on the PC.

Use the dropdown box to select the COM port that says *Prolific USB-to-Serial Comm Port or HRI-200 Communications device A.*

Figure 11: Wires-X configuration - Com Port

If the radio 'Prolific driver' or HRI-200 is not listed, click Refresh and try again. If that fails, click the Device Manager button, and select the COM Port listing. A COM port should disappear if you remove the cable and pop back into existence when you plug in the cable. Take note of the number.

Select the 'Serial port number specify' radio button, enter the COM port number into the box, and click OK. An 'invalid device' indication means you have to reinstall the device driver. I did that several times before I finally got it to work. The software kept choosing a COM port number that was already in use, which was not helpful.

Next, you will probably get the Wires ID Activation screen. A visible serial number indicates that the software can 'see' the connected device. Enter the Node and Room ID numbers that you received in the registration email from Wires-X.

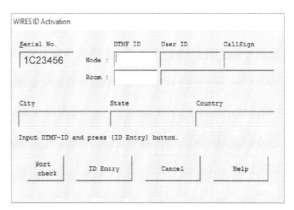

Figure 12: Wires-X configuration - ID activation

The **port check** only applies to the HRI-200 installation. You probably will not be able to enter your Node and Room ID numbers until it has been successfully completed. The 'COM port setting' popup checked that the software was able to communicate with the connected radio or HRI-200. This phase checks that the HRI-200 box can communicate with the Wires-X server.

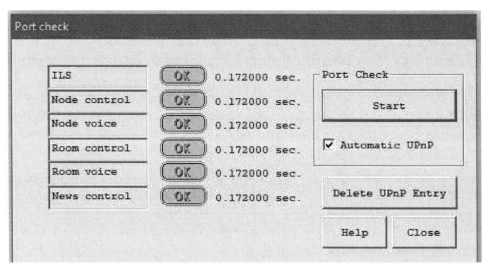

Figure 13: Wires-X configuration settings - port check

This should be easy. Just check the **Automatic UPnP** box and '**Start.**' If everything comes up with green 'OK', you are "good to go." Press **Close** and you should now be able to enter the Node and Room ID numbers that you received in your Wires-X registration email.

This is where my installation failed. Only the 'Node control' line was tested successfully. Even though I have UPnP activated both here and in my fibre router configuration, I had to 'open' the UDP ports. I should have been able to open the whole range of ports, but this was not successful. I ended up 'forwarding' each port individually.

To do this you need to access your main router. Refer to its handbook about the username and password required to access it and port forwarding. It will often have an IP address of 192.168.1.254. Well, mine does. If you can't find the logon, try 'admin' and 'password.'

1. You need to know the PC's Internal Host IP address. This is available in Windows Settings > Network & Internet > Ethernet - Properties (button). Look for the IPv4 address and write it down with the dots.

2. Enter the 192.168.1.254 or similar IP address of your main router into the URL box on your web browser. Access the router setup using the login and password from the manual.

3. The UDP ports that need to be forwarded are 46100, 46110, 46112, 46114, 46120, and 46122. If you plan to use the remote monitor program, you will also have to enable TCP port 46190. Try the remote monitor out first and only add the TCP port forwarding if it is needed.

4. In the **Admin > NAT > Port mapping** section, you should be able to add new port forwarding mapping. Click **New**.

    a. Interface: will default to your internet or fibre provider

    b. Protocol: Set to UDP or TCP/UDP, (not TCP)

    c. Remote Host: Leave blank

    You should be able to enter the whole range between the start port and the end port, but it did not work for me.

    d. External start Port: 46100

    e. External end port: 46100

    f. Internal host: The IP address you found in step 1

    g. Internal port: 46100

    h. The mapping name can be anything. I used 'Wires100,' 'Wires110' etc.

    i. Save the config. On my router 'Save' is labelled **Submit**.

5. Repeat step 4 for the rest of the UDP ports.

6. Some routers will not permanently save the configuration unless you reboot the router at this point. The configuration will work, but if you power down the router at some time, the configuration would be lost.

Wires-X software window

At this stage, you should be able to start the main Wires-X software window. It is huge. Some boxes should start filling up with data. There are a few more configuration settings to complete.

1. I found at this stage that the HRI-200 audio CODEC had hi-jacked my soundcard and turned off my normal audio speakers. Grrr! Click the speaker icon at the bottom of your Windows screen (or in settings). Click the sound slider and you should hear a bong. If not, use the dropdown to reselect your speakers.

2. In **File > WIRES ID information** you can see your ID data, which you can change if you want to. But you probably don't. You can set the information that comes up on your node display, in the text box where it says **Comment**.

I put my website there. Checking the **Confidential ID** box hides your callsign from the Node list. I didn't select it.

3. Check the **QSL exchange** box and you can add a 320x240 pixel .bmp image for the QSL information that appears when you connect to a node. Perhaps a club logo with text on it. There is room for a short bio or other text in the **Get Info** text box.

If you select 'Show position data' on the next page, your location will also appear on the QSL window.

Figure 14: Wires-X QSL information

4. You should enter your location on the **File > Location settings** tab. If a radio has been connected in the PDN mode, you can click **Read GPS Data** and the position will be updated with your current location. Check **Show position data** so that your location will appear on the Node Information QSL window. This also makes your location visible to stations that you call.

The next menu that needs attention is the **File > Property > Call settings** tab. I am not sure what a "round room" is. I think that it is a room that is public. It seems that you need the **Round QSO Room** checkbox checked if you want to be able to connect to a Wires-X room.

The **second check box** selects whether you will disconnect from the Wires-X room if someone links to your Wires-X node. If a friend might call you, it is probably best that you check this option. I have no friends, so I don't need to. **Back to Round QSO** will return you to the Wires-X room that you were connected to after you have chatted with the person who called your node. **Return to room** sets a default room. You can set a **Return room cancel** command. The only option I have set to 'checked' is the **Round QSO Room** connection.

TIP: The Help menu on the Wires-X software is context-sensitive. You will be taken to a help page that is relevant to the screen you are on. It is very helpful, and unlike the main program, it even has a search function.

5. **File > Property > General settings.** The only setting I changed was to extend the TOT (time-out timer) to 15 minutes. It disconnects the Wires-X room after the specified period. You can turn this off by checking the Unlimited TOT checkbox, but I prefer to have it turned on so the node will drop out when I go to bed. When the remaining time gets below 10 minutes a countdown timer is displayed below the radio ID at the top of the main screen. The **announcement** messages only relate to FM operation and will be greyed out if you don't have an FM transceiver connected. **Remote control** sets a password for access via the 'remote monitor software.'

6. **File > Property > Digital ID settings.** Enabling this transmits your ID signal at a preset interval from the donor radio. If you have a second 'Preset search' radio connected to an HRI-200 you can set this for that radio as well. This will send your ID to a repeater and probably send your GPS location. It is not active if the software is linked to a Wires-X room and the Test button will be inactive. I do not see any advantage in activating this feature, so I do not have the option selected.

7. **File > Property > Make list file.** This screen is used to save a copy of the current Group list, Active ID list, or Room ID list. Either as an HTML table which could be opened with MS Word or a web browser or as a .csv file which could be opened with MS Excel or a different spreadsheet program. I don't know why you would want to save these lists… but you can.

8. **File > Property > Log file settings.** This screen is used to set the name and activate logs of node activity and news feeds. I am yet to see any news on any node so that one is probably a bust. I have no interest in saving the node activity either, but I guess if you owned a busy room, it could be interesting.

WHEW! Are we done yet? Nope. One more to do.

9. If you are using the HRI-200 or the PDN method in the access mode with two radios, you need to setup the link to the handheld or access radio. Select **File > Transceiver (T)**. On the left are the voice channel settings. The area on the right is for the second 'Preset Search CH' radio that could be connected to an HRI-200. Yes! That would mean three radios just to access Wires-X. I am going to ignore that possibility. You would have to be very keen on Wires-X to even contemplate that option.

TIP: You don't have to fill in the Transceiver page if you plan on using one radio in PDN mode with no access radio.

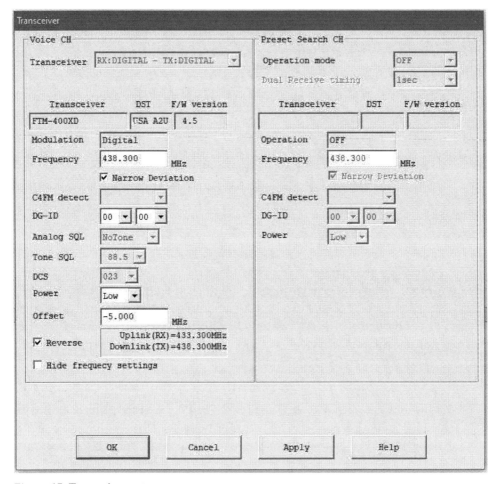

Figure 15: Transceiver setup page

*TIP: If your radio is compatible with the HRI mode, some data will be pre-populated. If it is not, you can enter the data. Use the **Help** button for more details*

Settings for a connected FTM-400XD

Transceiver: pre-populated and grey. It says that the donor radio is a digital radio.

Transceiver: pre-populated with the donor radio model.

DST: pre-populated for the donor radio. It is the radio's region information.

F/W version: Firmware revision, pre-populated for the donor radio

Modulation: Digital, pre-populated for the donor radio

Frequency: Enter the frequency that the donor radio will transmit to the access radio. Set the access radio to receive this frequency.

Narrow Deviation: I suspect this is only for FM. Selecting the VW mode on the access radio automatically changes to the DN mode as soon as you connect to the donor radio, and the donor radio sits on DN and will not allow you to change to VW mode.

C4FM detect: on my setup, this is blank and grey.

DG-ID: this is set at the default value of TX 0, RX 0. You should be able to change the DG-ID in the software and on the access radio, to improve privacy. People would not be able to hear the link between the donor and access radio. I tried it, but I was unable to get the access radio to connect.

Analogue squelch, Tone squelch, tone squelch frequency and DCS tone frequency are all for FM and disabled for digital.

Power is the RF power setting for the donor radio. Set it to low if you are using the radio as a hotspot. You could set high power if the donor is attached to an outside antenna, and you wanted to use the node across town. The same applies to the access radio. You would normally use the lowest power setting, but you could turn up the power if you are trying to access the node at a distance.

Offset: The offset depends on the frequency you have chosen. Bear in mind that the donor radio is transmitting a far higher power than a typical hotspot. It is likely to be able to be received at a considerable distance. So, the choice of frequency is important. It may be illegal in some areas to set up with a repeater offset. I chose an unused repeater slot and applied the correct – 5MHz offset for our band plan. Set the offset to zero, if you are using a simplex frequency out of the repeater sub-band.

The frequencies will be displayed below the offset. Remember to check (tick) the Reverse box. This is very important.

Hide frequency settings: Hides the RF frequency that you are using. Since 90% of the people on Wires-X are not accessing the network through a Fusion repeater, I think that it is better to hide the frequency. It is not very relevant in a hotspot situation such as PDN or HRI-200 operation.

That's it! The Wires-X software configuration is done! All you need to do now is setup the access radio if you are using one and learn about operating the software.

SETTING UP THE DONOR RADIO

There is no setup required on the donor radio. All the settings are done in the Transceiver tab of the Wires-X software – just discussed. However, you do have to set the radio into the Wires-X mode. Turn the radio off. Hold down the DX and GM keys while you press the power button to turn the radio on. The radio will start in the Wires-X mode. To return the radio to normal, repeat the process.

SETTING UP THE ACCESS RADIO

The setup for the access radio is the same as when you use it on a repeater or a hotspot. Any C4FM radio can be used. I am using my FTM-400XDR as the donor radio, so I will cover the two access radios I have. Other radios should be similar.

1. Set the radio to receive the frequency that will be transmitted by the donor radio. It is the frequency you set in the frequency box of the software.

   TIP: The donor radio will display its receive frequency and the access radio will display its receive frequency. So, if you have applied an offset, they will display different frequencies. When either radio transmits, they will both display the same frequency.

2. Set the offset to the offset you used in the Wires-X software setup.

3. Set the DN mode, TG-ID TX 0 RX 0

4. Save the channel with a name like 'Wires-X'

5. If you are using an HRI-200 you will hear any Wires-X node or room that has been selected in the Wires-X software.

6. If you have connected the radio directly in the PDN mode, the display will show either the donor radio's receive frequency or 'DIRECT.' A frequency display indicates that you are using the access point method with an access radio. The DIRECT indication indicates that you are using the direct terminal mode where you use the donor radio microphone and listen to the calls on the donor radio speaker. Press the lower **Dial** knob to swap between direct and access point modes.

   TIP: in the direct terminal mode all connections to Wires-X rooms must be made using the Wires-X software. The only live functions on the donor radio are the volume knob and pressing the lower Dial knob. In the access point mode, room search and select can be done on the access radio or the Wires-X software.

7. This step is optional. You do not have to enter the Wires-X mode if you are not intending to select Wires-X rooms on the radio.

   A long press of the **GM/X** button on an FT5D, FT3D, FT2D or pressing **F** **>** **AMS** on an FT-70D connects the access radio to Wires X. You can search for rooms and select rooms the same as you would when connected to a Fusion repeater. If you change the room using the Wires-X software, the change will be shown on the access radio's display.

OPERATING THE WIRES-X SOFTWARE

One issue with the software is that although there is a huge list of the Wires-X rooms and another huge list of the Wires-X nodes, there is no 'context sensitive search' function to shorten the list. In fact, there is no 'search' function at all. I was rather surprised at this omission.

Selecting a room

Find a room on the Wires-X list (bottom left) or the bookmarks list (top left). Right-click and select **Connect**. You can also select (highlight) a line on the list and click the Connect menu item at the top of the screen. There is a disconnect option there as well.

TIP: You cannot change to another room while a station is talking. You must wait for a gap in the conversation. You also cannot connect to a room if a conversation is in progress. Try again after a few seconds and hope to connect during a break in the conversation. This is the same as when you are using a repeater or a hotspot.

Status window

The software opens many windows and is best run on a large monitor. At the top right there is a status block which shows the connection to the HRI-200 or the donor radio. **Local** turns green when the access radio transmits. Or the donor radio in the 'PDN Direct' mode. **Net Digital** will be shown when you are linked to a room. Idle Digital means you are not linked to a room. The yellow box shows the connected room. On-air lock means the transmit is locked. Often because the donor radio is turned off. Click it to unlock. The grey box with -:- - shows a countdown timer if there are less than 10 minutes before the room disconnects. The TOT (time-out timer) is set on the Settings > Property > General Settings page.

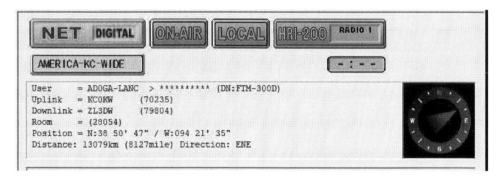

Figure 16: Wires-X status window

Below that there is information about the incoming call including, the callsign and message "FTM-300D," the room code, and their position, distance and direction from your location.

Room members

The room member's window floats above the rest of the screen and as far as I know, it cannot be docked. It contains the 'User ID' of people in the Wires-X room. The software picks an ID for you, but you can change it using Settings > Personal settings > Wires ID information > User ID change. Your User ID is your callsign and a suffix. ND means a PDN node, RPT means a repeater, and GW means a gateway. Your callsign should be on the list unless you opted to check the 'Confidential ID' box. The green cells, top left, indicate that JQ1LOI is talking on America Link.

Room ----AMERICA-LINK(21080) member 118 nodes					
AMERICALNK	JQ1LOI	Send Node : AMERICALNK (11080) / Mobile : JQ1LOI			
WM4RB-RPT	W6BML-RPT	K7REW-ND	W3QV-RPT	W5FC-RPT	KW5TX-TX
VE5BBZ-RPT	JF3KNW-PDN	VK5HDT-ND1	JAPAN-ND	GB7IS-RPT	K3PDR
WA3VHL-RPT	KC5OUR-GW	WA4TAW-ND	K3ERM-ND	W7ECA-RPT	PE1MVS-ND

Figure 17: Wires-X room members

QSL window

The QSL window pops up when you connect to a room. Often it contains no useful information, but in this case, it is quite detailed. If you are using an HRI-200 you can set the picture and details for your allocated Wires-X room.

Figure 18: Wires-X QSL window

The Wires-X room list

On the left at the bottom is the full list of Wires-X rooms. Yaesu calls it the 'active room ID list.' There is no search function, but you can change the ascending or descending index order by clicking on the column headers. **Act** is activity. It shows the number of people currently in the room. The vast majority of rooms are empty.

The active node ID list

Above the list of Wires-X rooms is the list of active nodes. This includes everyone currently logged in to the network via the Wires-X software. There is a list of what the coloured icons mean in the help system. Blue means the node is idle, and yellow indicates a network state, i.e. the node is receiving audio. You can connect to some nodes if they are not already connected somewhere else.

The group window

The group window is the top window on the left side. It is often obscured by the lavender-coloured room members list, but you can move that out of the way. There are three different options. The most useful is the Bookmark list.

Select **View > Group Window > Bookmark list** to display your favourite rooms. Add to the list by right-clicking a room on the Wires-X room list and selecting **Add Bookmark list**. The **Act** column shows the number of people connected to the room.

Select **View > Group Window > Connecting Node list** to show more details about the stations connected to the currently selected room.

Select **View > Group Window > MyRoom Access** list to show nodes connected to your Wires-X room.

Another way to add a room to the bookmark list is to use the menu system **View > Bookmark Info > Add > the DTMF room ID code.**

G.User ID	+DT...	Act	Call/Rm...	City	State	Cou...
----AMERICA-LINK	21080	117	America ...	Beau...	Texas	USA
----------CQ-UK	27793	018	CQ-UK	ABER...	Aberdee...	UK
AMERICA-KC-WIDE	28054	050	America'...	KAN...	Missouri	USA
ZL-MOUNT-GREY-NZ	69115	002	South Isl...	Chris...	Canterb...	Ne...
ADELAIDE-GW-10G	69159	010	VK5-GAT...	Adel...	South A...	Aus...

Figure 19: Wires-X bookmark list

Other windows

Below the status window, there is a running log of connection data and room activity. Below that there is a window which is for you to enter short 'chat' messages. They are useful during nets or just as a comment on the conversation. Click News to see if any news is there to read. I have never seen anything. The GM button seems to be a shortcut to some of the menu system windows.

Help system

The help system is very informative. It is well worth checking out.

HRI-200

This chapter is about configuring and using the HRI-200 for Wires-X. The HRI-200 is an interface that sits between a compatible Yaesu radio and your computer. This used to be the only way you could access Wires-X if you were not in the coverage area of a Yaesu repeater, but recent radios include the PDN and HRI modes which connect directly to your PC without the need for the HRI-200 box.

I will leave the question of "is it worth buying an HRI interface?" up to you. But I will lay out the facts as I see them.

1. The HRI-200 is the only way that you can host a Wires-X room. In fact, you get allocated a room number automatically when you register with Wires-X. You don't have to activate the room if you don't want to.

2. To use the HRI-200 you need a 'donor' radio. The combination of a Yaesu C4FM 'donor' radio and an HRI-200 interface makes a very expensive hotspot. In my opinion, it is unbelievable that the HRI-200 has no RF capability like every cheap hotspot does. There is also an option which requires two 'donor' radios and an 'access' radio. Three Yaesu radios in all.

3. You cannot talk over the donor radio. It is only used as a hotspot. You need a second radio, usually a handheld, to access the donor radio. The 'access' radio has full connectivity to the Wires-X network including connecting to Wires-X rooms and use of the news messages. Although I have never seen a Wires-X room where anyone has used the 'News' function.

4. You must use the Wires-X or 'Wires-X remote monitor' software to access the network. The donor radio cannot be used to make connections to nodes or rooms. The software **must** be running for the Wires-X node to work.

5. The Wires-X software supports the HRI-200 and the PDN/HRI 'terminal' modes.

6. Several Yaesu C4FM radios include the PDN and HRI 'terminal' modes which connect directly to your PC without the need for the HRI-200 box. That gives you access to the Wires-X network without a second Yaesu radio and without the cost of buying the HRI-200.

7. Because of the way Fusion works, you can use an FM radio as the donor radio or as a second connected radio. FM radios use DTMF tones for signalling and not all features are available. But you can talk to C4FM digital users and FM users via the Wires-X network.

COMPATIBLE RADIOS

The FTM-400XD and FTM-300D are the only C4FM radios, available new, that can connect to an HRI-200. The best thing to do would be to buy a used FTM-100D or similar. I bought an FTM-400XDR, but I am not contemplating leaving it permanently tethered to the HRI-200 as a Wires-X node.

The full list of compatible radios is, FTM-400XD, FTM-300D, FTM-200D, and FTM-100D C4FM radios. The FTM-6000 is listed as a compatible FM radio and other FM radios with a 6-pin or 8-pin mini-DIN data jack can be used. The list of compatible radios includes the R (international) and E (European) models.

HRI-200 FIRMWARE UPGRADE

The Wires-X node owner's page includes a firmware upgrade to the HRI-200. I don't know if it is necessary. I had so much trouble getting the program to recognise that the HRI-200 was connected, that I ran the firmware upgrade out of desperation. The program prompts you and it went quite smoothly, but my general advice is if the unit is working, you probably don't need to upgrade the firmware.

To do a firmware upgrade, download and extract the WIRES-X connection kit HRI-200 Firmware (Ver.1.01) zip file. Run HRI-200_ver101(ENG).exe and follow the instructions. Disconnect the HRI-200 USB cable. Unscrew the upper four case screws and loosen the lower two screws on the front only, to remove the top cover. Slide the small switch in the centre of the board to the PRG position. Plug in the USB cable to power the unit and load the new firmware. It is very fast. Then disconnect the HRI-200 USB cable, return the switch to normal, replace the top cover and screws, and reconnect the USB cable.

RADIO FIRMWARE UPGRADE

The HRI-200 manual suggests that you should do a firmware upgrade on your radio before using the HRI-200. This probably isn't necessary if you just purchased a new radio, but I had huge problems with the ADMS software, so I had already done a firmware upgrade on my FTM-400XDR. See page 25.

HRI-200 CONNECTIONS

The HRI-200 comes with a USB cable for the connection to your PC, and two data cables. One is a 10-pin cable used for connecting a digital radio (usually radio one) and the other is used for connecting an FM radio (usually radio two). Plug in the cable to the radio and then the cable to the PC. Then start the Wires-X software. There is an audio output connector. It uses a stereo plug, but it is only a mono signal. Start the donor radio in Wires-X mode. On an FTM-400XD hold down **DX** and **GM** Then use the **power** button to turn on the radio. Use the same process to exit the Wires X mode.

The PDN and HRI modes

The PDN and HRI modes can be used to access the Wires-X network in 'terminal' or 'access point' mode by connecting compatible radios via a USB cable to your computer. You need to register with Wires-X to get an ID and node number for the Wires-X software.

It was difficult to get the Wires-X software to work with the HRI-200, but it was easy to switch to the PDN mode once it was working. I have not tried the HRI mode, because it offers no benefits, and you guessed it! It needs a different interface cable.

PDN MODE

The PDN (portable digital node) method supports communication with other C4FM digital radios. It can be used in a 'direct' terminal mode using one radio that is connected via a data cable to your PC, or in 'access point' mode where a handheld radio is used to talk through the connected radio. The access mode gives you the freedom to move around the house and surrounding area depending on the antenna configuration and power of the doner radio.

PORTABLE HRI MODE

The HRI mode supports communication with FM and C4FM digital radios. It also supports a 'direct' terminal mode and an 'access point' mode that requires two radios. The only real difference between the PDN mode and the 'Portable' HRI mode is that it supports both C4FM digital and FM transmissions, so it requires an audio cable in addition to the SCU-19 or SCU-20 data cable. The SCU-39 and SCU-40 kits include the data cable and the audio cable. I have not experimented with the HRI mode, so I have not included it in the book.

COMPATIBLE RADIOS

These are the radios that support the PDN/HRI terminal modes and the data cables that are required.

Radio	PDN mode	HRI mode
FT5D	SCU-19 data cable	SCU-39 kit
FT3D	SCU-19 data cable	SCU-39 kit
FT2D	SCU-19 data cable	SCU-39 kit
FTM-400D	SCU-20 data cable	SCU-40 kit
FTM-400XD	SCU-20 data cable	SCU-40 kit
FTM-300D	SCU-40 kit	SCU-40 kit
FTM-200D	SCU-40 kit	SCU-40 kit
FTM-100D	SCU-20 data cable	SCU-40 kit

PDN MODE CONNECTIONS

Connect the 'donor' radio to a USB port on your computer with the SCU-19 or SCU-20 cable. Start the Wires-X software. If a COM port selection window pops up. See 'running the software' on page 51.

Set up the frequency for the access radio in the File > Transceiver menu of the Wires-X software. I used a pair of repeater frequencies that are unused in my area, but you can set a simplex frequency. Bear in mind that your donor radio may be transmitting as much as 50 Watts into an outside antenna. It will have quite a large coverage area, probably several miles or kilometres in radius. This may be greater if you have an elevated location such as an apartment building.

START THE DONOR RADIO IN WIRES-X MODE

Start the donor radio in Wires-X mode. On an FTM-400XD hold down DX and GM Then use the power button to turn on the radio. Exiting the Wires X mode is the same process. The display will show 'Yaesu Wires-X.' Once the Wires-X software is running and linked to the radio, the display will change to show the access frequency in access mode, or DIRECT in direct mode. I like the way you can switch between PDN access and the 'direct' terminal mode simply by pressing the lower Dial knob. In direct mode, you will hear incoming calls on the radio and be able to transmit back to them. The only other live control is the volume knob.

ACCESS MODE VS DIRECT MODE

In access mode, you will use your handheld or mobile radio to transmit through the donor radio to the network. You can search and link to rooms from the radio as discussed in the 'Using Wires-X' chapter or create a link using the Wires-X software as discussed in the 'Wires-X software' chapter. In the direct mode, you can only establish links using the Wires-X software. The big advantage of the direct mode is that you don't have to own two Yaesu digital radios. The big disadvantage is that you are tethered to the PC running the Wires-X software, by the USB cable.

IS WIRES-X BETTER THAN THE REFLECTORS?

Both networks work well. I think that it is a pity they both exist because it splits the Yaesu operators into two camps. I expect that is why the rooms that are bridged to reflectors are the most popular. Of course, this is a chicken and egg situation, because only the most popular Wires-X rooms have been bridged to reflectors. Having a local Fusion repeater is a good incentive to get onto the Wires-X network and using the PDN mode means you don't have to buy an MMDVM hotspot. Running a 50-watt node into an outside antenna will give you Wires-X access over quite a big area, perhaps a farm or the region around your amateur radio clubroom. If you already own a hotspot, there are many more reflectors than Wires-X rooms, and you gain the opportunity to talk with DMR, P25, and D-Star users.

BlueDV and Peanut

BLUEDV

BlueDV by PA7LIM is available in versions for Linux, Windows, Android, and Apple iOS. It is an application that is used without a radio, although it does need a DV dongle. It also works with some hotspots which need a serial connection to your PC or a Bluetooth connection to your phone. You can listen to three digital voice modes at the same time. However, I found the audio on YFS was a bit intermittent when I tried it. I believe that it is better to switch off D-Star and DMR while you are having a conversation on YSF.

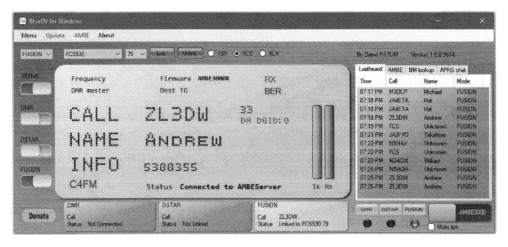

Figure 20: Blue DV by David PA7LIM (Windows version)

BlueDV will work with the following DV dongles and hotspots,

- Portable AMBE Server https://reflectorloversclub.jimdofree.com/

- the NWDR ThumbDV, (currently out of stock)
  https://nwdigitalradio.com/products/thumbdv%E2%84%A2)

- the DVMEGA DVStick 30 and Globetrotter, see https://www.dvmega.nl/

- the BlueStack hotspot
  https://www.combitronics.nl/index.php?route=product/category&path=62
  and the

- ZUMspot AMBE Server, see
  https://www.hamradio.com/detail.cfm?pid=H0-017021

  The ZUMspot AMBE Server has an AMBE3000 chip on board. It can be used with BlueDV, Peanut, DummyRepeater, or Buster software.

Links to software

http://www.pa7lim.nl/bluedv-windows/

http://www.pa7lim.nl/peanut/

https://github.com/g4klx/DummyRepeater

https://apps.apple.com/us/app/buster/id1060175273?mt=12

BLUEDV SETUP FOR YCS REFLECTORS

BlueDV works with the DG-ID mode where you can set static modules on the YCS network. It is similar to setting static talk groups on DMR. There are 20 different YCS servers, most are accessible via FCS reflector numbers. For example, if you set the **FCS** radio-button on the line just below the menu toolbar. Then select **Fusion > FCS530 >** and module number **37**. You will connect to module 37 (America Link) on the YCS530 reflector.

TIP: *you can find the valid options by clicking* **DG-ID List** *on any YCS reflector dashboard.*

Figure 21: Selecting an FCS reflector and module in BlueDV

If you create the link and transmit for a few seconds, you should see your callsign come up on the relevant YCS dashboard. Find the correct dashboard in the left panel at http://xreflector.net/. If you connected to YCS530 you could check out the dashboard at http://ycs530.xreflector.net/ycs/ and see your callsign listed.

The module you connected to will be displayed in the DG-ID column. If you have set static modules, they will be listed in the DG-ID column. The 79(10) means that I transmitted on module 79 which is not one of my static modules. It will time out after 10 minutes.

Nr.	Repeater	Name	QRG	ID	DG-ID
33	FCS53079 -ZL3DW (05)	Hamshack	434.3000	4029	05 35 37 79(10) 80 84 89

Figure 22: YCS dashboard showing static DG-ID modules selected

If you want to operate the unique arrangement where you will hear calls on several modules, you need to make an easy change to the **BlueDVconfig.ini** file. Search your 'My Documents\BlueDV' folder for it. Edit the file using Notepad or similar. Find the section headed [FUSION]. Add a new line with the modules you want to make 'static' on BlueDV. For example, **Options=5,35,37,80,84,89**. I selected Oceania, CQ-UK, America Link, NZ Canterbury repeater, and IPSC ZL DMR+ TG530. This will set those module numbers whenever you link to any YCS reflector.

Set the startup reflector to FCS

In BlueDV, turn off **Serial**, and select **Menu > Setup**.

Set **Enable at start** so that Fusion is enabled when the program starts. Make the default reflector **FCS** and set the FCS dropdown box to an **FCS reflector and module**. In this case FCS530 module 37, which is how you access YCS530 module 37.

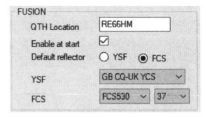

Once you have set up static module numbers, they will be active whenever you connect to any YCS reflector. You connect via an FCS reflector that is running on the YCS system. This can have unforeseen consequences because the DG-ID list on the reflectors varies. For example, YCS530 module 80 is the Mt Grey repeater, YCS310 module 80 is the QuadNet array, and YCS235 module 80 is FreeStar. So, although the module numbers remain static, the group you are listening to may be completely different. Some modules are common to most YCS reflectors. DG-ID 2 to 7 are almost always the same.

Go to the YCS reflector dashboard and select **DG-ID-List** to check out the modules for your YCS reflector. The YCS reflectors are listed on http://xreflector.net/.

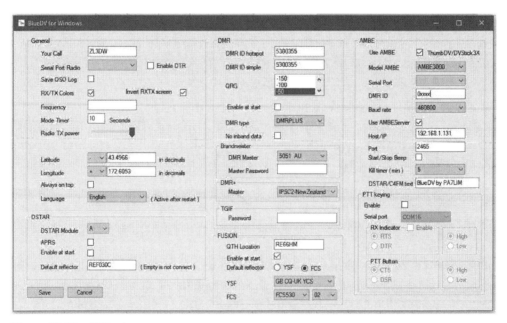

Figure 23: Blue DV setup screen

DV DONGLES

I have experimented with two Digital Voice dongles. They allow you to access the DV modes without a radio. In most cases the software of choice is BlueDV.

ThumbDV AMBE 3000 USB stick with BlueDV

The Thumb DV by www.NWdigitalRadio.com looks similar to the DVMEGA DVStick 30, but the board layout and chipset are different. I purchased a pre-owned one from TradeMe, our local online auction site. The dongle will work with a PC running BlueDV, or an Android phone running the BlueDV AMBE (Beta) software. I had some difficulty getting started. The software hardly ever recognised the dongle.

I reloaded the USB driver from https://ftdichip.com/drivers/vcp-drivers/ and at the same time, I added a YSF repeater in the BlueDV setup screen. The BlueDV website recommends this if you  have problems. I don't know which action fixed the problem. But after doing both, the software works fine. Turn **Serial** on. If the top line of the main BlueDV display says **Firmware AMBE3000R** and you can operate the DMR, D-Star, and/or Fusion switches, the connection is working.

*TIP: If the receive audio is distorted, update the USB driver. You must use the **460800** baud rate (except for very early ThumbDV dongles). If you have connection problems, and you have already updated the USB driver, look up the Com port in Windows Device Manager and set the Port Settings to **460800, 8, None, 1, None**.*

To use the Android phone option, download and install the BlueDV AMBE (Beta) software for the phone from the Google store. The dongle is plugged into your PC which has to be running the AMBE Server utility, https://www.pa7lim.nl/garage/. BlueDV AMBE is not available for the iPhone.

Portable AMBE Server

The Portable AMBE Server is another DV dongle that can get you onto YSF reflectors without using a Yaesu radio. It is available from the XRF reflector lovers club in Japan. The unit contains a Raspberry Pi Zero W and a modem board containing the AMBE 3000 vocoder chip. It supports D-Star, C4M (YSF), and DMR. The unit can be powered by a USB power supply or a USB port on your PC. The power connector is the micro USB port closest to the short end of the case and the LED.

There are three ways to use the AMBE Server dongle. The normal method is to connect the unit to your home WiFi and use it from within your home network with BlueDV running on a computer or BlueDV AMBE (Beta) running on an Android phone. BlueDV AMBE is not available for the iPhone.

The second option is to use the 'tethered mode,' which involves configuring your Android phone to act as a WiFi hotspot and connecting the AMBE Server dongle to it. This lets you use the dongle anywhere there is cell phone coverage.

The third and most difficult method is to plug the unit into a USB port on your PC and run the AMBE Server utility, available at https://www.pa7lim.nl/garage/. In this scenario, you can access the AMBE server from an Android phone anywhere in the world. The phone connects to the AMBE server at home via the cell phone system or any connected WiFi such as a hotel or restaurant free WiFi service. You will be able to make D-Star, YSF, or DMR calls from anywhere. I have not tried the tethered or remote options because I don't have an Android phone. There are setup videos for the normal WiFi and tethered modes on the website.

Setting up the Portable AMBE Server for local WiFi was quite straightforward.

1. Remove the four black case screws only
2. Remove the SD card from the Raspberry Pi. It just slides out – no latch.
3. Insert the SD card into a card reader and plug that into a USB port on your computer.
4. Very important! Ignore all the Windows error messages and do NOT format the SD card. Close each Windows error message that pops up.
5. Windows Explorer should pop up with the SD card 'boot' drive directory
6. Run **AMBEDconfig.exe**. Change the SSID and password message to your WiFi ID and password. Input the AMBE Server IP address 192.168.xx.131/24. It is labelled wlan0 on the box. The first three numbers should be the same as the WiFi router's IP address (often 192.168.1). I had to change the third number from 43 to 1, so I ended up with 192.168.1.131/24.
7. WiFi router's IP address, e.g. 192.168.1.8. It should be in your phone's WiFi setup information. Click **Save**.
8. Remove the SD card and replace it into the Raspberry Pi. Power up the AMBE Server dongle. Open BlueDV setup. Check Use AMBE and Use AMBE Server. Set the Host IP to the AMBE Server IP address (192.168.1.131). **Save > Serial**.

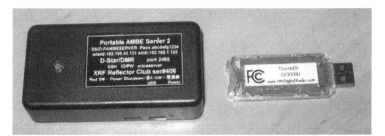

Figure 24: Portable AMBE Server and Thumb DV dongles

PEANUT

Peanut is another PA7LIM creation. It is a phone app for Android phones that supports connection to various D-Star reflectors and DMR talk groups. There is also a Windows PC version.

The Peanut project provides access to 39 YSF reflectors, plus D-Star XREF and XLX reflectors, and DMR talk groups. There are also 'Peanut' rooms which are not part of any other digital voice network. It does not have direct links to FCS reflectors, but there are plenty of choices. You don't need a dongle or a digital voice radio. The phone or PC does the routing to the network.

To use Peanut, you need a 'Peanut' code from https://register.peanut.network/. You do not have to be registered with Yaesu to access the Yaesu reflectors, but you do have to be D-Star registered to access the DPlus (D-Star) reflectors and registered for DMR if you want to access DMR talk groups. To do that you must have an amateur radio callsign. Some of the reflectors are transcoded so that you can talk between digital voice modes.

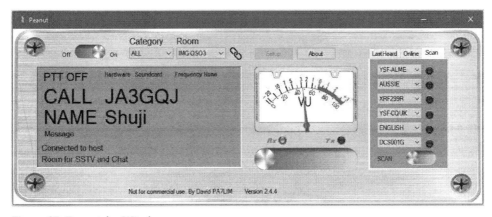

Figure 25: Peanut for Windows

You can select your six favourite talk groups, or reflectors and scan them for activity.

There is a list of Peanut rooms at http://peanut.pa7lim.nl/rooms.html

The Peanut dashboards are at,

- Global http://peanut.pa7lim.nl

- Japan http://peanut.xreflector-jp.org

- USA http://peanut-usa.pa7lim.nl

Radio programming software

YAESU ADMS SOFTWARE

I found the Yaesu programming software, or more particularly the interface between the radios and my computer to be absolutely terrible. Hopefully, I am a rarity, and you will have no problems. I spent many hours trying to get the Yaesu ADMS programs to work.

The cable that is supplied with the FT5D only works for firmware updates. You need a different cable for programming or the Wires-X PDN mode. The cable that is supplied with the FT-70D works for both firmware and program updates. But the procedure is very odd and unreliable. The FTM-400XD was supplied with the SCU-20 programming cable.

Each radio has a different version of the programming software, and each requires a different interface cable. The ADMS software is available from the Yaesu website. Find your radio model and look under 'Software.'

FT-70D ADMS-10

I had a lot of difficulties getting the ADMS-10 programming software to interface with the radio. I had to download the firmware upgrade and run the USB driver program to even get Windows to recognise the radio without throwing a 'device not recognised' error. After that, it did create a COM port, but I got a Timeout error every time I asked the program to download data from the radio. It also refused to upload data to the radio. The button was greyed out.

I had fewer problems when I loaded the software onto a laptop PC, but it was still a nightmare. At that stage, I knew that it was possible to make the software work, so I went back to my PC and tried again. In desperation, I carried out a firmware update despite the radio not needing one. It made no difference. The programming software still refused to talk to the radio. Finally, I downloaded the USB driver for the fourth or fifth time and Bingo! It worked.

The ADMS-10 software is so bad that neither the supplied manual nor the downloaded advanced manual include any mention of it.

The process required in the manual is just plain ridiculous. No other manufacturer requires anything like the gymnastics required by Yaesu. I will explain the process, even though it does not often work (at least for me).

1. Start the ADMS-10 software and click the upload icon (arrow pointing to a computer). Do not click OK yet.

2. Remove the battery from the radio

3. Plug in the external power supply

4. Turn the radio on

5. While the radio is turned on, unplug the external power supply

6. Plug in the supplied USB cable between the radio micro USB and the PC

7. Hold down the AMS button and reconnect the external power supply

8. ADMS will be displayed on the FT-70D

9. Click OK on the computer screen

10. Quickly press the **band** button on the radio. If you are lucky the data will transfer. It sometimes works, but I nearly always get a Time Out error. Sometimes it helps if you press the **band** button and then click OK on the computer.

On other radios, this would be simple! Turn off the radio, attach the USB cable, hold down a nominated button, and start the radio. Then initiate the data transfer.

When you save a memory channel in the FT-70D ADMS-10 software the Dig/Analog' choices are **Analog** or **DN** with a separate column to set **AMS** on or off. You can save a channel with the **VW** mode selected if the VW voice-wide option is enabled in settings.

FT5D ADMS-14

The process for the FT5D is better. You do not have to remove the battery, the process is more logical, and there are fewer steps.

1. Download and run the ADMS-14 software. Click the upload icon (arrow pointing to a computer). Do not click OK yet.

2. Connect the SCU-19 data cable, (not the USB cable that comes with the radio).

3. Turn the radio on while holding down the F-MENU key.

4. The FT5D display will show CLONE

5. Click OK on the computer screen

6. Quickly press SEND on the FT5D and the data will transfer.

The kicker is that you have to buy an SCU-19 cable. Or the SCU-39 kit which includes the SCU-19 cable. You can substitute the CT-169 cable if your PC is old enough to still have an RS-232 serial port and is powered by steam. This is an expensive radio. In my opinion, the SCU-19 programming cable should be supplied.

What I can't work out is why the supplied cable, which carries the firmware update data from the computer to the radio just fine, cannot be used for programming the radio. It makes no sense! The only explanation I can think of is that it is a way of making you pay Yaesu more money to buy a cable you probably only need once. It has certainly put me off buying any more Yaesu radios.

The ADMS-14 software is so bad that neither the supplied manual nor the downloaded advanced manual include any mention of it.

You can set a channel to **DN**, **AMS**, or **FM**. AMS is auto mode select. And you can save a channel with the **VW** mode selected if the VW voice-wide option is enabled in 'Settings.'

If you don't have the cable, you can use the program by writing the configuration data to a micro SD card and then placing the SD card into the radio and importing the revised configuration. It is a bit tedious and clumsy, but it works. You will probably want to buy an SD card for the radio anyway.

TIP: You need the SCU-39 kit if you want to use the WIRES-X Portable Digital Node (PDN) function. Does the FTM-400XD use the same cable as the FT5D? No, of course not!

FTM-400XD ADMS-7

Yet another version of the ADMS software and a different programming cable. However, I am pleased to report that the required SCU-20 cable was supplied with the radio. The ADMS-7 software is so bad that the supplied manual makes no mention of it.

Before attempting to use the programming software, save the current radio configuration to the SD card. Select **SD Card > Backup > Write to SD >All > OK > Back > Back > Back**. This is important because if the data transfer fails the radio will reset to factory defaults.

Weirdly the FTM-400XD software only allows you to save channels as FM, AM or NFM. There are no digital options. It appears that the radio never saves the mode into the memory channels. It just transmits on whatever mode you set with the **DX** button. When you turn on the radio, the upper Band A display will come up on the mode that was in use when you turned the radio off. The lower Band B display will always come up in FM because it does not support digital. This could be annoying if you wanted to save a mix of FM and digital channels to the upper display. It is probably better to have only digital channels on the Band A display and only FM (or AM air band) channels on the Band B display.

The radio has separate memory banks, programmable memory search (PMS) banks, home channels, and VFO data for the Band A display and the Band B display.

The settings menu has 'Normal' settings, and tabs for GM_Wires-X digital, APRS, and APRS beacon modes.

No	Receive Frequency	Transmit Frequency	Offset Frequency	Offset Direction	Operating Mode	Name	Tone Mode
1	438.12500	433.12500	5.00000	-RPT	FM	Hotspot	OFF
2	438.40000	433.40000	5.00000	-RPT	FM	ZL_YSF	OFF
3	439.57500	434.57500	5.00000	-RPT	FM	ZL3DVR	OFF
4	432.75000	432.75000	5.00000	OFF	FM	DV_Simpl	OFF
5	147.05000	147.65000	0.60000	+RPT	FM	75	OFF

Figure 26: ADMS-7 memory channels

Uploading the radio configuration data

I had problems with the FTM-400XD ADMS-7 software. It would upload the configuration from the radio but then throw a Windows memory error after the upload was completed. It would not display the contents of the radio. I tried loading the config from the SD card, but the only option was **ALL**. The **Memory** and **Setup** options were greyed out. You guessed it, the ALL option loads but fails. So, at this stage, the clone cable method fails, and the SD card option also fails. "Way to go" Yaesu! I eventually fixed the problem using the following process.

The SCU-20 method

Try to read the radio configuration using the SCU-20 cable. You will probably find that you cannot write to the radio without reading from it first. With luck, this will work, and you won't have to resort to the SD card method.

1. Save the current radio configuration to the SD card. Select **SD Card > Backup > Write to SD >All > OK > Back > Back > Back.**

2. Connect the SCU-20 cable to the radio and a USB port on your computer.

3. Start the ADMS-7 software. Click the upload icon (arrow pointing at a computer), but don't click OK. The instructions below are displayed.

4. On the radio hold **DISP/Setup** to get to the main menu.

5. Select **Reset/Clone > Clone > This radio -> other**

6. Click **OK** on the ADMS software and immediately touch, **OK?** on the radio.

7. With luck, you will see the data transfer and then 'completed' on the radio and the software. The screen should now show the channels that were stored in the radio. If you keep getting Windows errors, proceed to the SD card method.

8. If the data transfer is successful, you can add memory channels and change the radio settings.

9. When you have finished editing you can save the file on your PC and download it back to the radio. This time you use the download icon (arrow pointing to a radio), and **Clone > Other this radio** on the radio. The timing can be tricky. If you get it wrong it will factory reset the radio, "nice!" I find it best to click, **OK?** on the radio and then **OK** on the computer.

The SD card method

When I tried to use the SD card method, I found that although I could use the **ALL** function when I saved the radio configuration to the SD card, I was not able to use the **ALL** option to read the SD card with the ADMS-7 software.

This was a major problem because the option to read the individual 'Memory' and 'Setup' backup files separately was greyed out and unavailable. "Stalemate," I thought.

However, after I had tried unsuccessfully to read the radio configuration using the SCU-20 cable, the individual Memory and Setup files option became available, and I eventually got the data from the radio into the software. I sent that data back to the radio over the Clone cable. After that, I can use the clone cable successfully to read the radio configuration and save it back to the radio. And it only took about 12 hours to figure it out.

So, this is what worked for me.

1. You need an SD card in the radio.

2. Press and hold the **Disp/Setup** button.

3. Save the current radio configuration to the SD card. Select **SD Card > Backup > Write to SD >All > OK > Back > Back > Back.**

4. Remove the SD card from the radio, and place it into your SD card reader (see page 102). Plug the reader into a USB port on your PC. Use Windows Explorer to find the drive.

5. Start the ADMS-7 software. Select **Communications > Get data from SD card > All.** Look for the data file in FTM400D\Backup\Clone.

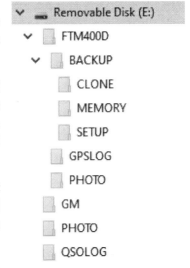

6. Try to upload it to the ADMS. If it loads without an Error that's it! You can proceed with updating the settings and memory channels. Save it back to the SD card using **Communications > Send data to SD card > All.**

7. Replace the SD card in the radio, then recover the new data using **SD Card > Backup > Read from SD >All > OK > Back > Back > Back**.

8. Or you could try the clone cable method described above.

9. If you get an error in step 6, you can upload the individual memory and settings files from the SD card.

10. Select **Communications > Gct data from SD card > MEMORY**. Look for the data file in FTM400D\Backup\Memory. If the option is greyed out, try the cable clone method as discussed in the second paragraph of this section. That made the option available when I went through this misery.

11. Select **Communications > Get data from SD card > SETUP**. Look for the data file in FTM400D\Backup\Setup. If the option is greyed out, try the cable clone method as discussed in the second paragraph of this section. That made the option available when I went through this misery.

12. If successful, you can proceed with updating the settings and memory channels. Save it back to the SD card and use **Communications > Send data to SD card > All**.

13. Replace the SD card in the radio, then recover the new data using **SD Card > Backup > Read from SD >All > OK > Back > Back > Back**. Or you could try the clone cable method described above.

RT SYSTEMS SOFTWARE

According to many posts on the Yaesu forums, the RT Systems software works very well. I have used their software for other radios, but I have not tried the Yaesu versions. They are a good company to work with and they have a great reputation for quality and excellent support.

I noticed that for the FT-70D they abandon the connection technique favoured by Yaesu and use the clone mode instead. Their cable connects to the speaker/mic jack instead of the micro USB jack. There is a video on their website that demonstrates the Yaesu ADMS method working intermittently but failing most of the time. So, I'm not the only one who has had trouble with it.
https://www.youtube.com/watch?v=RRy-TO588sk

The problem remains that there is a separate program for each of my three radios, at US $25 each and I need two additional data cables at US $30 each. The USB-57B 'clone' programming cable for the FT-70D and the USB-68 cable for the FT5D. There is a discount if you buy the cables bundled with the software. But it would still cost me US $125 for my three radios, for software and cables that should be supplied for free with the radios.

Yaesu radio SD cards

The FTM-400XD and FT5D transceivers have a micro SD card slot for an optional SD card. The SD card is not supplied with the radio, but it is worth buying one. It holds vital backup information, and it can be used with the ADMS software or to setup a similar radio.

The SD card holds

- A backup of the memory channels (option)
- A backup of the radio settings (option)
- GP-ID (group ID) settings
- GPS logging
- Storing photos taken with the optional MH-85A11U microphone
- Storing data sent using the GM function or Wires-X

You can use any micro SDHC card from 2 Gb to 32 Gb. Don't pay extra for a very fast card. It is not necessary. It is unlikely that you will be able to buy a micro SD card smaller than 32 Gb.

Yaesu recommends 'initialising' the new SD card. I never bother and have not had any problems, but it is your risk. Hold DISP > SD CARD > FORMAT > OK?

Yaesu recommends turning the power off the radio before inserting or removing an SD card. I never bother and have not had any problems, but it is your risk.

FTM-400XD

The micro SD card slot is on the body of the radio on the same side as the microphone and remote head cables. It is inserted with the copper strips facing the same side as the speaker. i.e. down if the radio is mounted in a vehicle or up if the radio is sitting on your workbench. It will only go in one way. Don't force it. The card will latch when it has been installed. A white SD icon is displayed in the top right part of the display when a card is installed. A flashing icon indicates that data is being read or written.

Saving a backup

I strongly advise saving the radio setup and memory channels to the SD card. If anything goes wrong with programming using the ADMS software the radio will reset to a blank state. The same can happen with a power glitch, or according to Yaesu, "incorrect operation, static electricity, or electrical noise."

Hold DISP > SD CARD > BACKUP > Write to SD > ALL > OK?

This saves three files. The CLNFTM400D.dat clone file which the ADMS program fails to read, the memory channels MEMFTM400D.dat, and settings SYSFTM400D.dat.

Recovering a backup

If the radio resets and asks for your callsign you can recover your saved memory channels and setup data from your saved files. You cannot proceed until you enter a callsign or just a single letter. Then **Hold DISP > SD CARD > BACKUP > Read from SD > ALL > OK?** The radio should reboot automatically, and everything should return to normal. You might notice minor changes like the mode on Band A being different to the way you left it because that is not saved in the setup.

FT5D

The micro SD card slot is above the external DC input under the rubber protection strip. Insert the micro-SD card with the copper strips facing the front of the radio.

The SD card method works well with the ADMS-14 software, which I can't use with the clone method because I don't have the required SCU-19 cable. It looks like the ADMS software can only import the full backup file, not the memory channel file.

The folder structure is self-explanatory. The FT5D\BACKUP folder contains the backup file, both memory channels and settings. The FT5D_MEMORY-CH folder has a backup of the memory channels, assuming that you saved them separately.

Saving a backup

I strongly advise saving the radio setup and memory channels to the SD card. If anything goes wrong with programming using the ADMS software the radio will reset to a blank state. To save a full backup use,
F MENU > SD CARD > BACKUP > Write to SD > F MENU > OK > OK

To save the memory channels only,
F MENU > SD CARD > MEMORY CH > Write to SD > F MENU > OK > OK

TIP: 'Waiting' means writing or reading to the card. Wait for the job to be 'completed.'

Recovering a backup

If the radio resets and asks for your callsign you can recover your saved memory channels and setup data from your saved files. You cannot proceed until you enter a callsign or just a single letter. Then **Hold DISP > SD CARD > BACKUP > Read from SD > ALL > OK?** The radio should reboot automatically, and everything should return to normal.

F MENU > SD CARD > BACKUP > Read from SD > F MENU > OK > OK
F MENU > SD CARD > MEMORY CH > Read from SD > F MENU > OK > OK

GPS location

The FTM-400XD, FTM-400D, FTM-100D, FT5D, FT2D, FT1XD, FT1D, FT-991A and FT-991 can send latitude and longitude data. The FT-991 can only send a manually entered position because it does not have a GPS receiver. The FT-70D does not have a GPS receiver and it can't send a manually entered position, but it can display the position of a received caller if the text is included in the received message string.

FT5D

The GPS receiver should be turned on by default. As well as relaying your position to other stations it keeps the time display accurate.

GPS information screens. Press **F menu > Disp**. Touching the screen cycles between two displays. Press the BACK key or PTT to exit.

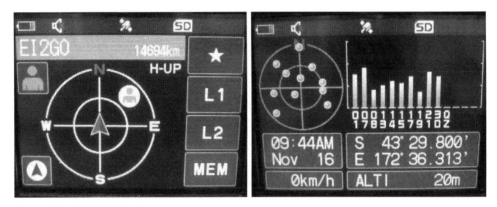

Figure 27: FT5D GPS information screens

Touch the blue person icon to display the distance and bearing to the last received callsign. You can save it my touching MEM and L1 or L2. There is a menu setting to change from the compass display to a numeric display. In that version touching the up arrow indicates your location.

Touch the centre of the screen to change to the second GPS display. It shows the time and date calibrated by the GPS signal, your latitude, longitude, speed and altitude (often inaccurate). The image at the top left is a 'radar' plot of the received GPS satellites. The outer ring is the horizon, the inner ring represents an elevation of 45 degrees, and the centre is directly overhead at an elevation of 90 degrees. Blue dots indicate satellites with valid data, the grey dots are satellites where the signal strength is too low. This is usually because they are just about to set over the horizon, or they are just coming into range. The numbers indicate that the radio is receiving the GPS constellation (numbers 1-32). Ten satellites are about the maximum that will be visible from my location.

The signal bar labelled QZ means the GPS receiver is receiving one of the Japanese QZSS satellites. They are only available in Japan, parts of Asia, some of Australia, and sometimes New Zealand.

FTM-400XD

The GPS receiver is turned on by default. As well as relaying your position to other stations it keeps the time accurate.

Pressing the DISP button cycles through the compass display, time and date, and this radar image which shows the received GPS satellites. See the note about this display in the FT5D section.

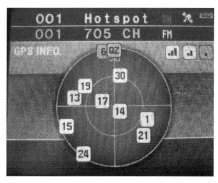

Figure 28: FTM-400XD GPS satellites

The satellites are in the GPS constellation which uses numbers between 1 and 32. Grey numbers have low received signal strength and will not be used for calculating your location.

Clock

I like the clock. It has analogue and digital 12-hour or 24-hour displays, plus the date. Touching the mode icon reveals a lap timer and a countdown timer. It is a shame it cannot display local and UTC, given that there are two clock displays.

APRS

APRS beacons can only be sent from FM channels using the 'B band' display. Since that is for FM only APRS is out of scope for this book. There is a guide to the APRS mode on the Yaesu downloads page. Look for FT5DR_DE_APRS_ENG_2108-A.pdf.

APRS.FI WEBSITE

Some reflectors will pass your position on to the APRS.fi website https://aprs.fi Enter your callsign followed by a star, e.g. ZL3DW* into the Track Callsign search box. I have had success with YSF27793. A short 2-3 second "kerchunk" is all that is required.

Hotspot APRS Host setting

Turning on **APRS Host Enable** lets the hotspot announce its position on APRS.fi. As far as I know, it only sends the position when it boots up, or when it resets. My hotspot resets every night. It also resets if you click **Apply Changes**. I leave APRS turned off since my hotpot is not travelling anywhere. This setting sends the location you entered in the Pi-Star setup screen, not the location of your radio. Select the **APRS Host** that is closest to your location, or the rotating host, **rotate.aprs2.net**.

Pi-Star Dashboard

This chapter covers the features of the Pi-Star dashboard after you have configured it using the instructions in the next few chapters. There is a lot of information presented on the main dashboard page. Open Pi-Star on your web browser and click **Dashboard**. Try http://pi-star/ or http://pi-star.local/

*TIP: many folks think that Pi-Star is running on their PC. It is actually running on the Raspberry Pi board. The dashboard is a web page used to control Pi-Star. You can access the dashboard from any web browser on any computer that can access your local LAN over Ethernet or WiFi. For example, it runs fine on my iPhone. Enter **pi-star.local** into the URL entry box on the browser.*

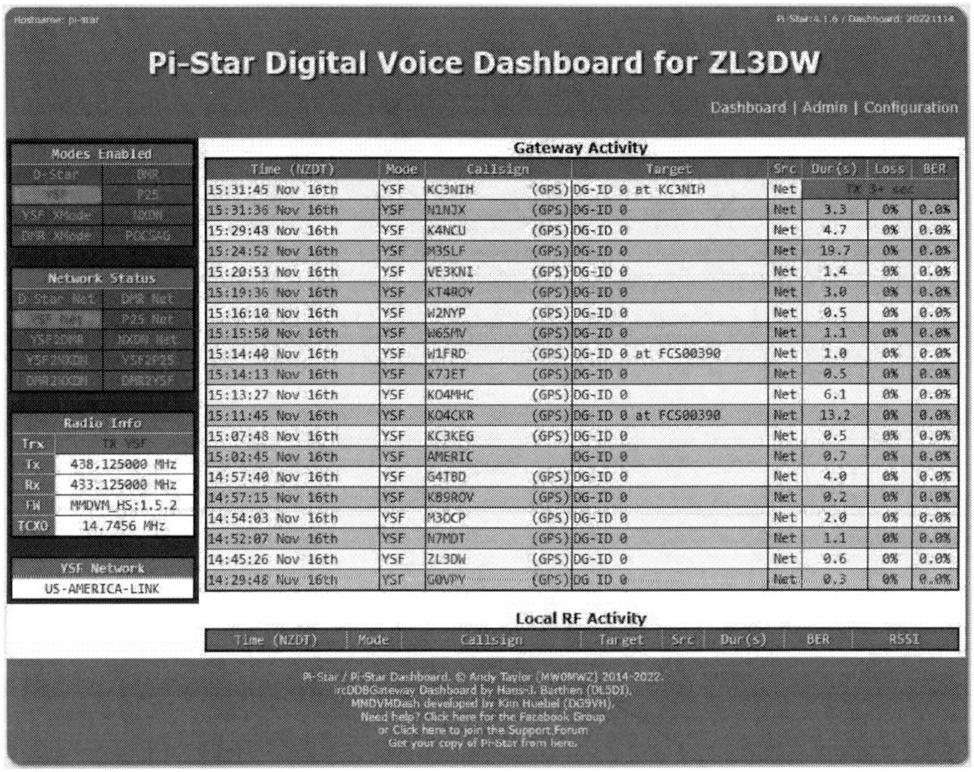

Figure 29: Pi-Star Dashboard page

Modes Enabled shows the modes that you selected in the Configuration screen. This book only covers the YSF and YSF cross-mode options.

Network Status. YSF Net (green) indicates that the internet connection is up, and the system is communicating with a YSF network. If you are using cross-mode options you may see green YSF2DMR, YSF2P25, YSF2NXDN, or DMR2YSF indicators.

Radio Info shows the status of the hotspot.

Trx

- Listening (green) the hotspot is in its idle condition waiting for a call to arrive from the network or the RF path.

- Listening YSF (yellow) means that the user has stopped transmitting and the hotspot is waiting for the network hang time to expire. If no signal is heard during the hang time the hotspot will go to its idle condition.

- TX YSF (red), means that the hotspot is transmitting to your access radio.

TX and **RX** show the hotspot's transmit and receive frequencies, set in the configuration.

FW is the firmware revision of the hotspot.

TCXO is the nominal oscillator frequency. This is what you adjust if you apply a frequency correction offset. They usually run at either 12.288 MHz or 14.7456 MHz.

YSF Network shows the name of a connected reflector, or 'Not Linked.'

GATEWAY ACTIVITY

The gateway activity area shows calls broadcast by the hotspot that originated on the internet (network) side. These are the calls indicated with 'Net' in the Src (source) column. Calls with 'RF' in the source column originated as an RF signal received by the hotspot. Usually, that will only be your handheld or mobile radio(s). If the Pi-Star dashboard is for a repeater, there could be many RF calls listed.

Other information includes the date and time of the call, the mode, the callsign, the target DG-ID that was called, the source, the duration of the call, the amount of data loss percentage (after forward error correction) and the BER (bit error rate) quality of the signal. Clicking on the callsign takes you to the user's nominated website. Usually but not always their QRZ.com listing. Note that (GPS) does not mean that the radio is equipped with GPS. It is a link to the APRS.fi website with the callsign loaded as the search quantity. A green bar indicates the station that is calling.

LOCAL RF ACTIVITY

The local RF activity area provides pretty much the same information for calls that originated on the RF side of the hotspot. Plus, the received signal strength indication (RSSI). For a hotspot on your desk, it is usually massive, S9 +46dB (-47 dBm). The RSSI indication is more useful for measuring the signal from stations accessing a Pi-Star based repeater.

THE ADMIN PAGE

The 'Admin' page adds the ability to link to a reflector using the **YSF Link Manager**. The access username is pi-star. The password is raspberry. You can select from a list of YFS and FCS reflectors. YCS reflectors are usually accessed through FCS reflector links. Select **Link** and **Request Change**. Unlink the same way. Note that you cannot change reflectors or unlink while there is a conversation on the reflector. Click the dropdown and enter text into the search box for a context-sensitive search. For example, entering 'America' highlights all reflectors with America in the name. 'FCS' jumps to the start of the FCS list. 'YCS5' finds the Australian and New Zealand YCS reflectors. Try 'CQ,' Quad,' or 'UK.'

At the top of the page, there is **Gateway Hardware Information** about the Raspberry Pi. It shows the Host Name (which we decided not to change), The Pi Kernel release, The Pi model, its CPU loading per core, and the CPU temperature. If it gets into 'the red' you need to add a heatsink or fan to improve the Raspberry Pi CPU cooling. Mine runs in the orange zone at 52.5 degrees Celsius, but it seems to be stable, and I do have a heatsink on it.

I don't know much about the **Service Status** section. I guess, "green is good." MMDVMHost means the hotspot is working as an MMDVM host. YSFGateway is the link to the YSF network. YSFParrot is a link to the parrot server, for test calls. The other green icon is PiStar-Watchdog which indicates the status of the Pi-Star software. If there is a problem the timer will elapse, and the indicator might go red. It is more likely that the hotspot will just freeze up.

Even though it is the 'administration' page, the reflector linking is the only setting you can change. But you do get three more menu items at the top of the screen. **Live Logs** records incoming and outgoing calls, **Power** to shut down or reboot the Pi, and **Update**. WARNING: only click this if you want to do an update. It takes several minutes, and you won't get asked again!

THE CONFIGURATION PAGE

If you are configuring a Pi-Star hotspot from scratch, check out the 'Hotspots,' 'Loading Pi-Star,' and 'Pi-Star Configuration' chapters. This section covers the configuration of the hotspot <u>after</u> you have it working and connected to your local network.

The configuration page adds some more menu items at the top of the page. **Expert** for experts like us, **Power** to shut down or reboot the Pi, **Update**, only click this if you want to do an update. You won't get asked again! **Backup/Restore**, and **Factory Reset**. Do not click that or you will erase all your configuration settings. You probably will not need to access any of the Expert pages. It is better to leave them alone unless you really are an expert. You do have to change one expert setting to set up the EA7EE version of Pi-Star.

Power

The **power** menu item provides this attractive image. Click the green **reboot** icon to restart the hotspot or click the red **shutdown** icon to safely shut down the hotspot. The file system on the Raspberry Pi SD card can become corrupted if you just pull the power out. It is always best to shut down the Pi using a button on the hotspot display if it has one (mine does not), or this

power button. I admit that I usually just pull the power plug, but that is a risk that I take, and I know how to restore the software if I have to.

Backup/Restore

You should make a backup as soon as your configuration is stable and working. Backups are downloaded to your standard Windows download directory as a Zip file. You can leave them there or move them to another folder.

Nothing seems to happen when you select the 'download configuration' icon, but the program will copy a file to your Windows download directory.

You have to click **Choose File** before you click the 'restore configuration' icon. Otherwise, nothing happens.

The zip file contains twenty configuration files. Editing settings in the 'Expert' area changes the relevant config file. Most are text files that can be viewed or edited in Notepad or Wordpad if you are especially brave. I have copied text from an old backup file to correct a DMR network configuration error after I messed up the settings. You can restore the configuration from the Zip backup file if you want to revert to an older configuration or copy your hotspot configuration to a new hotspot.

Apply changes

You will see **Apply Changes** at the bottom of each section. It should be used every time you make changes in any of the configuration sections. New options often appear after you click **Apply Changes**. Click the button and wait for the hotspot to reboot, then check what changed before proceeding to the next section.

Control Software section

Control Software: select **MMDVM Host**. The repeater option is for a repeater controller card, so you probably won't use it.

Controller Mode: should be set to **Simplex Node** for a simplex hotspot, or **Duplex Repeater** for a duplex hotspot. I use a duplex hotspot.

TIP: There is no big advantage in using a duplex hotspot for YSF. There is a big advantage in using a duplex hotspot for DMR. It allows you to transmit on one time slot while receiving a signal on the other, or to receive two timeslots at the same time. A simplex hotspot will block you from transmitting if a call is being received in the other time slot.

YSF repeaters cannot transmit two voice signals at the same time, so a simplex hotspot works just as well as a duplex one. If you plan to use DMR sometime in the future, buy a duplex hotspot.

MMDVM Host Configuration section

This is where you select which mode the hotspot will work on. Select **YSF Mode** by clicking the box beside the label.

You can select other modes such as D-Star or DMR if you have radios for those modes.

Leave the hangtime settings at the default value of 20. It can be confusing when calls come in and you could miss part of a conversation on one mode if the hotspot switches to a different mode. The hang time settings create time (20ms) for someone to come back to your call before a call on another mode can take over.

If you are using multiple modes and this becomes a problem, you can increase the hangtime to alleviate the issue.

You have the option of turning on several cross-mode 'transcoding' options. I cover them in more detail near the end of this chapter.

- YSF2DMR lets you access the DMR networks using your Yaesu radio.

- YSF2NXDN lets you access the NXDN reflectors using your Yaesu radio.

- YSF2P25 lets you access the P25 reflectors using your Yaesu radio.

- DMR2YSF lets you access the YSF reflectors using a DMR radio.

The hotspot converts the data format from one coding system to the other. It is very clever indeed. I have tried all of the options and they work fine, but you can talk to other digital voice mode users via linked XLX reflectors without adding this layer of complexity. I have a DMR handheld radio, so I can just turn DMR on in the hotspot. There is no advantage in using the cross-mode option. I did quite like P25.

The **MMDVM Display Type** is set according to what display (if any) you have on your hotspot. See MMDVM Display on page 112.

If you have made any changes, click **Apply Changes**.

General configuration

The general configuration section contains information about the hotspot and your location. It remains the same no matter what mode you select, so it should already be configured. The hostname should be pi-star, I don't recommend changing it. The node callsign will usually be your callsign unless the Pi-Star configuration is for a public access repeater. If you select **Auto,** on the URL line the software will choose your QRZ listing address. You can select **Manual** and enter a different website if you want to. Again, this option is really for public access repeaters. Radio/modem type is critical. It must match your hotspot. But it should have been set up by now. See the Hotspots and Pi-Star chapters for setup details.

The Node Type should be **private** unless you have a good reason to make it public.

Only turn on **APRS Host Enable** if you want to send your position to the aprs.fi website. This makes your location public. The hotspot will send a beacon to aprs.fi whenever it is re-booted. If you leave it turned off, the Yaesu stations you work will get the usual distance and bearing position display but there will be no beacons sent to the aprs.fi website, unless the connected reflector forwards the data.

The APRS Host dropdown box provides a choice of 'rotating' APRS host sites. Choose one near you.

The time zone and language were set in the initial setup. Usually as a part of the Pi-Star download from pi-star.uk.

If you made any changes, click **Apply Changes.**

YSF configuration

The Yaesu System Fusion configuration section will be shown once you have selected the YSF mode on the MMDVM Host Configuration and clicked **Apply Changes.**

YSF Startup Host: can be set to start the hotspot linked to any YSF or FCS reflector. YFS41562 CQ-UK and YSF32592 America Link are the busiest by far.

Uppercase Host Files: enables a lowercase to uppercase conversion in the software. I believe that this is only necessary if you are using an FT-70D. But I leave it turned ON because you have to do a software update, not just the usual Apply Changes reboot if you change the setting. If you have an FT-70D make sure it is **ON,** if not it does not matter.

Wires-X pass-through: must be enabled so that you can select reflectors from the radio. Leave it turned **ON.**

Mobile GPS Configuration

This setting has nothing at all to do with the GPS receiver on the radio. It is used when you connect a GPS receiver dongle to the Raspberry Pi that is hosting the MMDVM hotspot. You can turn on the serial port or USB port if a USB dongle is plugged in and the baud rate. Normally 4800 or 9600 for a GPS receiver but may be faster for a dongle.

Firewall configuration

Firewall configuration is for experts. Leave all the access settings set to **Private** and Auto AP **on** and uPNP **on.**

Auto AP lets the Raspberry Pi act as a WiFi access point if it is unable to find your WiFi network within 120 seconds. You can link your phone to the node and configure the WiFi access on the Raspberry Pi. uPNP lets the Raspberry Pi manage its firewall settings. If you turn it off, you can configure the firewall settings in the Expert tab.

Setting Dashboard access to **public** would allow internet users to observe the Pi-Star dashboard. This is OK providing you changed the 'Remote Access Password' to stop them from changing the configuration settings. Making the ircDDBGateway public would allow others to manage routing. Changing SSH access enables remote users to access the Raspberry Pi via Secure Shell. This could be required if someone is offering remote support to fix a Raspberry Pi configuration issue. Note that SSH access within your LAN is always available. You can access the Raspberry Pi via the Pi-Star SSH screen or the puTTY SSH (secure shell) program without changing this setting.

Wireless Information and Statistics

Wireless Information and Statistics shows the WiFi connection. It should state 'Interface is Up.' You can configure, reset, or refresh the WiFi if it is not working.

Auto AP SSID

This setting changes the default so that a password and login name is required to access the hotspot in AP mode. If you need the AP mode, you are probably in enough trouble without adding this extra layer of complexity.

Remote Access Password

The remote access password should only be set if you are making the hotspot public. You would normally only do that if you were using Pi-Star on a repeater, and you wanted to make the dashboard available for repeater users.

If you are the only person that needs to see the hotspot dashboard, keep it set to private, and do not enter a remote access password.

YSF TRANSCODING MODES

Transcoding effectively does the same thing as the bridges between digital voice mode reflectors and talk groups. It converts the data you send to the network through the hotspot to the required format and then converts the signal received from the network back to the Yaesu format. Since all the modes use the AMBE Vocoder, this is not too difficult. Note that it is not changing the RF modulation, only the data coding. The MMDVM hotspots can't transcode D-Star, although I have heard that the OpenSpot 4 hotspot can.

I am not certain that these modes are worth the effort. You can use bridged reflectors to talk to users of the other digital voice modes more easily than configuring a transcoding mode. You cannot use YSF and the transcoder at the same time because the YSF host has to be set to YSF0002 for DMR, YSF0003 for NXDN, or YSF0004 for P25. You can change to another YSF or FCS reflector using the admin page, but that ends the transcoding process.

YSF2DMR

YSF2DMR lets you access the DMR networks using your Yaesu radio. You can use any of the DMR networks on the Pi-Star list. But not multiple networks, because you cannot set the DMR Master to 'DMR Gateway.' I tried the local DMR+ IPSC2-NZ-HOTSPOT server. I don't think you can set static talk groups, but I was able to key up a dynamic talk group with no problem. I also connected to BM_5051 Brandmeister Australia and was able to set static talk groups on the Brandmeister dashboard.

Note that you do require a DMR ID number to use this feature, and you must register with Brandmeister to use the Brandmeister network, TGIF to use their network etc. In fact, DMR+ is the only DMR network that only requires a DMR ID.

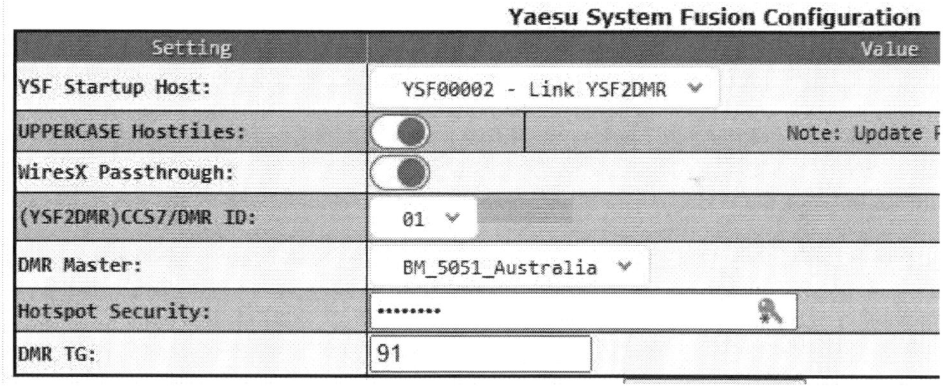

Yaesu System Fusion Configuration

Setting	Value
YSF Startup Host:	YSF00002 - Link YSF2DMR ⌄
UPPERCASE Hostfiles:	⬤ Note: Update F
WiresX Passthrough:	⬤
(YSF2DMR)CCS7/DMR ID:	01 ⌄
DMR Master:	BM_5051_Australia ⌄
Hotspot Security:	•••••••• 🔑
DMR TG:	91

Figure 30: YSF2DMR setup

Turn YSF2DMR **on** in the MMDVM Host Configuration zone and **Apply Changes**.

In the Yaesu System Configuration zone set

- YSF Startup Host: to **YSF00002 - Link YSF2DMR**.

- YSF2DMR DMR ID: should auto-populate with your DMR ID number. It can be added in the General Configuration zone if necessary.

- The image above shows that I am connected to BM_5051 Brandmeister Australia.

- Hotspot Security is your Brandmeister password. If you connect to TGIF, FreeDMR etc. it will be the password for that service.

- DMR TG: is the startup talk group. **91** 'Worldwide' is a good starting point because it is the busiest DMR talk group.

- **Apply Changes.**

The Pi-Star dashboard and admin page will show a YSF2DMR information box containing your DMR ID and the connected DMR server. Also, the YSF2DMR and YSF XMODE indicators will have turned green.

YSF2NXDN

YSF2NXDN lets you access the NXDN mode using your Yaesu radio. You have to register at https://www.radioid.net/ to get an NXDN ID number. There are only 6532 registered users, compared to 228,699 registered DMR users.

Turn YSF2NXDN **on** in the MMDVM Host Configuration zone and **Apply Changes.**

In the Yaesu System Configuration zone set

- YSF Startup Host: to **YSF00003 - Link YSF2NXDN.**

- Enter your NXDN ID number.

- Enter a startup host. There are 173 NXDN talk groups. Check out the list at https://www.pistar.uk/nxdn_reflectors.php

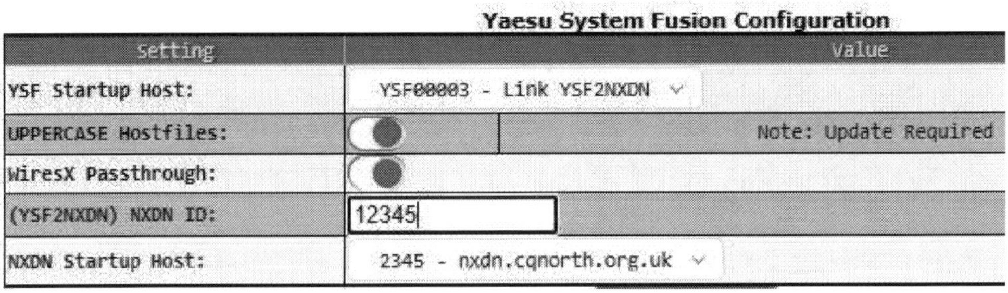

Yaesu System Fusion Configuration

setting	value
YSF Startup Host:	YSF00003 - Link YSF2NXDN ∨
UPPERCASE Hostfiles:	⬤ Note: Update Required
WiresX Passthrough:	⬤
(YSF2NXDN) NXDN ID:	12345
NXDN Startup Host:	2345 - nxdn.cqnorth.org.uk ∨

Figure 31: YSF2NXDN setup

The Pi-Star dashboard and admin page will show an NXDN Radio information box containing a RAN number and the connected NXDN talk group. Also, the YSF2NXDN and YSF XMODE indicators will have turned green.

NXDN (next-generation digital narrowband) is an open standard digital voice system developed by Icom and JVC Kenwood between 2003 and 2006. NXDN can be either 12.5 kHz or 6.25 kHz wide and can be configured as a dual-channel system with two 6.25 kHz voice channels occupying a 12.5 kHz channel slot. The system is not very popular for amateur radio and so far, it has failed to replace DMR, P25, and Tetra in the commercial world. NXDN uses talk groups rather than reflectors. A RAN number acts like the TG-ID on YSF. RAN 1 is used for talk groups, and sometimes a different RAN number is used for local conversations through the repeater.

YSF2P25

YSF2P25 lets you access the P25 reflectors using your Yaesu radio.

You must use the VW voice wide mode for P25.

Turn YSF2P25 **on** in the MMDVM Host Configuration zone and **Apply Changes**. In the Yaesu System Configuration zone set

- YSF Startup Host: to **YSF00004 - Link YSF2P25.**

- Enter your DMR ID number. P25 uses the DMR registration process.

- Enter a startup host. There are 222 P25 talk groups. Check out the list at https://www.pistar.uk/p25_reflectors.php

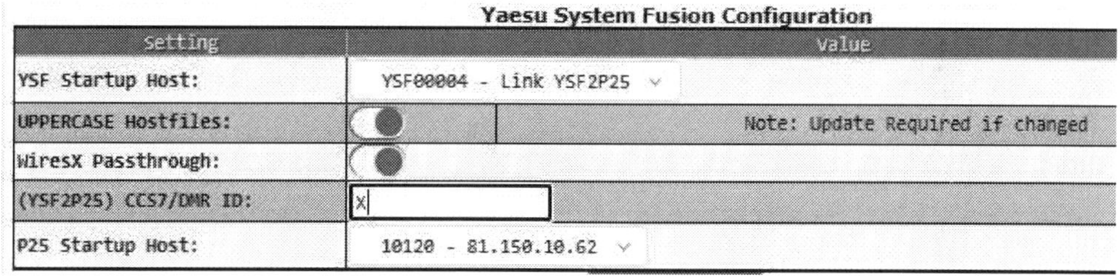

Yaesu System Fusion Configuration	
Setting	**Value**
YSF Startup Host:	YSF00004 - Link YSF2P25 ∨
UPPERCASE Hostfiles:	⬤ Note: Update Required if changed
WiresX Passthrough:	⬤
(YSF2P25) CCS7/DMR ID:	x|
P25 Startup Host:	10120 - 81.150.10.62 ∨

Figure 32: YSF2P25 setup

The Pi-Star dashboard and admin page will show a P25 Radio information box containing a NAC number and the connected P25 talk group. Also, the YSF2P25 and YSF XMODE indicators will have turned green.

Talk group	Region	Dashboard
TG 10100	Worldwide	m1geo.com/p25
TG 10200	North America	http://dvswitch.org/P25_NA/
TG 10201	North America TAC1	http://dvswitch.org/P25_NA_TAC1/
TG 10300	Europe	https://p25-eu.n18.de/
TG 10400	P25/NXDN Pacific	http://pacificp25.repeaters.info/
TG 302	Canada	http://p25canada.hopto.org/p25canada/
TG 10	Parrot	
TG 9999	Unlink	

Figure 33: P25 dashboards

Selecting a P25 talk group

You can change P25 reflectors from the radio. On the FT5D set the **VW** mode and connect using the **GM/X** button in the usual way. Search **ALL** or use the **ID-based search** to select a talk group. I was not able to connect the FTM-400XD in VW mode. It always reverts to DN, and the search function does not perform very well. There is a way around this.

Connect using the DX button. Hold the # key on the microphone until DTID and a # come up on the radio display. Use the microphone buttons to enter the talk group number and then press # again. The number must be 5 digits long, but you can pad it with leading zeros. To use the channel, you must leave the connected mode by pressing the DX key so that you can set the voice mode to VW.

The FT-70D does not have a search function, but you can change P25 talk groups on the radio. Connect using F AMS, rotate the Dial to En* and enter the talk group number, press AMS. The number must be 5 digits long, but you can pad it with leading zeros. You must leave the connected mode by holding down the MODE key. Then a short press of the MODE key changes the voice mode to VW and you can use the P25 channel.

Selecting a DMR or NXDN talk group

You can change DMR or NXDN talk groups from the radio. On the FT5D connect to the hotspot using the GM/X button in the usual way. Search ALL or use the ID-based search to select a talk group.

On the FTM-400XD connect using the DX button. You can search using the ALL or context-sensitive search just like looking for a YSF reflector. Or you can enter a TG number directly. Hold the # key on the microphone until DTID and a # symbol appear on the radio display. Use the microphone to enter the talk group number and then press # again. The number must be 5 digits long, but you can pad it with leading zeros.

The FT-70D does not have a search function, but you can change DMR or NXDN talk groups on the radio. Connect using F AMS, rotate the Dial to En* and enter the talk group number, press AMS. The number must be 5 digits long, but you can pad it with leading zeros.

DMR2YSF

DMR2YSF lets you access the YSF reflectors using a DMR radio. You cannot set DMR2YSF and YSF at the same time. Why would you use a DMR radio to access YSF reflectors if you have a YSF radio? But you can set DMR and DMR2YSF at the same time. It works in the same way as using two DMR networks on the hotspot at the same time. The YSF talk groups are prefixed with a 7 which the hotspot uses to tell that the data is destined for the YSF network, not the nominated DMR network. You have to program the YSF reflectors as talk groups on your radio one channel for every reflector, the same as programming DMR talk groups. The DMR master must be set to DMRGateway and I think that DMR network 1 must be Brandmeister. Considering that there are several DMR networks to choose from with thousands of talk groups, I cannot see much point in using a DMR radio to access the YSF network. But you can if you want to. There is an even less useful DMR2NXDN option.

Hotspots

If you have a local YSF or Fusion repeater and you are happy talking to users through that, you do not need a hotspot. However, if you do not have access to a local repeater, or you want the freedom to choose any reflector, at any time, a hotspot is the way to go. A word of warning though. Buying hotspots is a bit addictive. Next thing you will want two so you can use one for DMR or D-Star.

Most of the hotspots "out there" are MMDVM simplex hotspots. They transmit and receive on the same frequency. Most models can be used for all of the popular digital voice modes, YSF, DMR, D-Star, NXDN, POCSAG, and P25. But I am only going to talk about YSF. Most MMDVM hotspots use a Raspberry Pi Zero W or Raspberry Pi Zero 2W single-board computer to run the Pi-Star routing software. All Pi Zero hotspots could also run on a Raspberry Pi 3, or Pi 3 Model B, but the larger form factor does not match the size of the hotspot. Duplex hotspots are bigger and need more computing power, so a Raspberry Pi 3 or 3B is preferred. The Pi 3 supports 2.4 GHz and 5 GHz WiFi, but the Pi Zero only supports 2.4 GHz WiFi. Raspberry Pi Zero boards are much slower, particularly when booting up, but the gap is narrower with the new Pi Zero 2 which is five times faster than the old Pi Zero. Some hotspots use other single-board computers. Pi-Star.uk has ISO images for Raspberry Pi, Nano-Pi, Odroid XU4, and Orange-Pi Zero boards.

Duplex hotspots have two antennas and they run in duplex mode like a repeater.

Note that the Raspberry Pi 3 requires a bigger power supply (2.5A recommended, 700mA minimum) than a Pi Zero with an OLED screen. I have had no problems running off a powered USB 3 hub. The 2.4" Nextion screen draws 90 mA. You need an additional USB to TTL adapter to use Nextion screens with most simplex hotspots.

MMDVM RASPBERRY PI HOTSPOTS

There are dozens of different models which may come with cases or plastic protection. Most require you to buy a Raspberry Pi Zero or model 3 elsewhere. A search online will bring up many options. Here are a few, I found. Note I am not endorsing any model. I have not tried most of them.

TIP: It is often difficult to work out what you are going to get. Especially with the Chinese suppliers. For example, some websites show a picture of a hotspot with a Pi Zero. But they do not ship with a Pi Zero. The rule is. If the advert does not specifically say something is included… it is probably NOT included. Even if there is a photo showing the item. Some hotspots are supplied assembled, some are not. Some include a case, and some don't.

BI7JTA duplex hotspot. Unfortunately, the model I bought is no longer available. But similar models are available online. It is a duplex hotspot that was provided with a 2.4" Nextion screen, a programmed SD card, and the Raspberry Pi 3 model B.

https://www.bi7jta.org/cart/. Apparently, the cost of the STM32 chip has increased dramatically due to short supply.

TGIF Spot. https://tgifspot.com/ has a range of simplex hotspots. One with an OLED display, one with a 2.4" Nextion display, and one with a 3.5" Nextion display.

RFinder created the SkyBridge+ dual-band Simplex and the HCP-1 duplex hotspot. The HCP-1 includes an internal battery and is very portable. The computer is a Raspberry Pi Zero.

LZ duplex and LZ simplex hotspots, some come with a 3.2 inch Colour Screen. They are MMDVM hotspots designed for a Rpi Zero, (not supplied).

Jumbospot dual-band simplex MMDVM for Rpi Zero or 3B. These are a clone of the ZUMspot. Sold from a wide variety of vendors with a wide variety of prices depending on what you get. Most often they do not include the Raspberry Pi Zero, and sometimes they do not include the case.

ZUMspot https://www.hamradio.com/detail.cfm?pid=H0-016491. These are simplex UHF hotspots with a 1.3" OLED screen. They require you to buy a Raspberry Pi Zero or Zero 2. They will also run on an RPi 3 or an Odroid SBC.

Rugged Spot supplies a series of 'NEX-GEN' simplex MMDVM hotspots, some with Nextion screens and some with ceramic antennas (which work very well). They are supplied pre-programmed which is very nice. They are based around a JumboSpot supplied with a Raspberry Pi3-B and a plastic layer case. https://hamradio1.com/product/rugged-spot-store/

TIP: Don't worry if the hotspot is not assembled. All you will have to do is solder the RPi header pins and sometimes the RF SMA connectors (make sure they are supplied though). My simplex hotspot kit came with all the required screws and two short headers for the Raspberry Pi. The hotspot board was 100% complete. You will need an SD card and some free software to create the Pi-Star image and probably a micro USB cable or a USB power supply.

NO PI HOTSPOTS

Some hotspots are not based on the Raspberry Pi. A few have battery power and can be carried around in your pocket.

The SharkRF OpenSpot 4 has a built-in battery. It is a fully cased simplex hotspot with WiFi. The OpenSpot 4 uses propriety 'Shark RF Link' access software, not Pi-Star. It features a much faster bootup time than a typical MMDVM hotspot. Both models can do transcoding, but the PRO version can transcode D-Star to DMR, YSF or NXDN, which Pi-Star can't do. The OpenSpot hotspots are quite expensive, but worth considering because Raspberry Pi boards are hard to get at the moment.

https://www.sharkrf.com/products/openspot4/

SIMPLEX HOTSPOTS

Most people use simplex hotspots. They are cheaper, easier to set up, and faster to use than duplex hotspots. Simplex hotspots receive and transmit on the same frequency. Almost all simplex hotspots are MMDVM modem boards paired with Raspberry Pi Zero W or Zero 2W single-board computers.

The OpenSPOT4 by SharkRF is an exception. It is a self-contained simplex hotspot that can be slipped into your pocket and carried around. The OpenSPOT4 is not an MMDVM hotspot. Instead of Pi-Star, you use 'SharkRF link' for configuration using any web browser. Configuration and connection to a WiFi network seem to be very easy. The only downsides are, it can only connect to one DMR network at a time, and it is more expensive than a MMDVM simplex hotspot.

Raspberry Pi Zero 2W

You will probably have to buy a Raspberry Pi Zero to supply the computing power for your MMVM hotspot. They are not usually supplied with the hotspot. You must get the 'Zero W' version which has WiFi. I recommend getting the new Raspberry Pi Zero 2W. It is five times faster than the older Zero W board.

The SD card and display

If you are providing your own Raspberry Pi, you will also have to buy and prepare a micro SD card for the Raspberry Pi. It contains the Linux operating system for the Raspberry Pi and the Pi-Star modem software for the hotspot. I cover that in the 'Loading Pi-Star' chapter starting on page 101.

Because of the small form factor of the MMDVM modem board and the Raspberry Pi Zero, most simplex hotspots have a small 1.3" OLED display or no display at all. If you want the large display, you can connect a Nextion display to the Raspberry Pi Zero using a USB to TTL adapter. I wouldn't bother. The Nextion displays work better with a duplex hotspot and a Raspberry Pi 3B.

Building a simplex hotspot

Don't be scared of building a simplex hotspot. It is just an assembly job. The MMDVM modem board will probably be supplied 100% complete with no soldering required.

Some dealers offer a cheaper price if you solder the SMA RF connector and/or the header pins. You will probably have to solder in some header pins on the Raspberry Pi Zero anyway. No problem if you have a temperature-controlled soldering iron and at least some experience with soldering components onto a printed circuit board.

The header pins for my Raspberry Pi were supplied in the hotspot kit. Which was a nice bonus. It also included the case, screws, and spacers. I will step you through the process. It is easy.

The picture shows the kit that I received. The PCB was supplied in an anti-static bag and no soldering was required on the hotspot board. Just plug the hotspot board onto the Rpi.

Figure 34: A typical simplex hotspot kit. No soldering required.

The case clips together and it is a bit tricky to open. I wedged one side open with a small flat screwdriver and slipped a strip of cardboard into the gap. Then I applied the screwdriver to the other side and the case slid apart easily. Don't reassemble the case until the board has been mated with the RPi Zero and screwed onto the base section. And remember to remove the protective plastic from the screen by pulling the small tab, before the final assembly.

I chose to pair the MMDVM modem with the Raspberry Pi Zero 2W. They usually have no header pins installed, but I was lucky, and they were supplied in the hotspot kit. If the Raspberry Pi Zero has the full header row you can throw the supplied header pins away.

Solder the header pins into the rows at both ends of the board. Solder the short pins to the PCB. The long side should face up.

Figure 35: Raspberry Pi Zero 2W (no header pins)

Put both sets of pins into the board before flipping the board over, so that it sits flat when it is upside down for soldering. The plastic must be flat on the board so that the pins are exactly vertical.

This Pi Zero was supplied with header pins.

Figure 37: Raspberry Pi Zero W (with header pins)

Figure 36: Snip these four leads short if the RPi has a full row of header pins

Use side cutters to trim the four pins on the display board so they won't short out pins on the Raspberry Pi GPIO header. This is especially important if the Raspberry Pi has a full row of header pins.

Place the Raspberry Pi into the case making sure that the SD card slot is accessible. Insert the two short screws on the header side of the Raspberry pi board. Carefully stack the boards, using the supplied spacers on the side furthest from the header plugs. Insert the long screws through the hotspot modem board, the spacers and the Raspberry Pi and tighten, (not too tight).

You might prefer to power up and configure the board before installing the top case. After that, clip the top case on, being careful not to damage the micro USB and HDMI jacks on the side. Carefully ease the case around them and they will pop into the shell. If the board does not have a ceramic patch antenna, screw the antenna onto the SMA connector and power up the board. The micro USB closest to the short end of the board is the power connector.

Pi-Star notes for a simplex hotspot

The Pi-Star configuration is covered in the Loading Pi-Star and Pi-Star configuration chapters. There are some minor differences when using a simplex hotspot. Click **Apply Changes** regularly after each configuration step. It takes some time for the modem to reboot, but the configuration screen changes as a result of your previous choices. If something looks odd, such as two entry boxes for frequencies, click **Apply Changes** and the page will reformat.

- The controller mode is Simplex
- There is only one frequency
- The display option is usually OLED /dev/tty/AMA0
- The Radio / Modem type is different to a duplex modem

Figure 38: My MMDVM simplex hotspot

DUPLEX HOTSPOTS

More and more people are using duplex hotspots, especially to take advantage of the two timeslots offered by DMR. Other than the larger display with more information, there is no advantage in using a duplex hotspot for other digital voice technologies.

A duplex hotspot acts like a repeater, re-transmitting the signal that it receives. The circuit board is typically about twice the width of a simplex hotspot with a form factor designed for mounting on a Raspberry Pi 3 board rather than an RPi Zero. They have two RF chips and two antennas, and they use two frequencies. You can set any frequency split between receive and transmit, but I believe it is best to stick with the standard ±5 MHz offset on the 70cm band and ±600 kHz offset on the 2m band.

Your local band plan may use different offset splits. I chose a repeater pair that was not in use in my area. Or as luck would have it, anywhere in New Zealand.

Duplex Advantages

A duplex hotspot offers significant advantages over simplex hotspots if you are planning to use the hotspot for DMR. But apart from the much nicer display, there is no advantage for YSF or D-Star users.

Bigger display

My duplex hotspot is driving a Nextion display giving a nice bright colour display of the YSF activity. It is an MMDVM hotspot running on a Raspberry Pi Model 3B, and although I have tried the Pi-Star alternatives covered later, I am using the standard Pi-Star software. The displayed frequencies are hard coded into the display firmware, and they are usually incorrect.

Figure 39: My duplex hotspot showing a call from the USA

POWERING UP YOUR HOTSPOT

Make sure the antennas are plugged in. Some models have ceramic antennas on the board, but most have a small two small SMA antennas.

Plug the power cable into the Raspberry Pi. It uses a micro USB connector. You can power the Rpi Zero from a plug pack, a USB 3.0 port on your computer, a USB hub, or a USB power supply. A Raspberry Pi 3 is a bit more power-hungry than a Pi zero, especially if you have a Nextion display. Most phone chargers and USB 2.0 ports cannot supply enough power and the device may not work reliably. The official Raspberry Pi power supply can supply 2.5 amps at 5.1 volts. You may also need a 'micro USB' to 'USB type A' cable.

A Raspberry Pi 3 will boot quite quickly. If you have a Nextion display it will show the idle screen as soon as the power is applied. The screen pages are coded into the display, not loaded from the Raspberry Pi. It will take a minute or so for the WiFi to get established with a DHCP network address.

Depending on the screen firmware, the IP address might be listed at the bottom right of the idle screen. Eth0 means a wired Ethernet Connection and wlan0 indicates a WiFi connection.

A Raspberry Pi Zero will take a minute to boot, followed by some more time for the WiFi to get established with a DHCP network address.

TIP: for Simplex or WiFi-only hotspots. If it is the first boot and you copied the wpa_supplicant.conf into the SD card boot drive, the Rpi will boot, loud the WiFi config, and then reboot. It can take several minutes.

MY NEW HOTSPOT DOESN'T WORK

Don't panic! I covered this situation in the troubleshooting section, on page 124.

The most common problems are,

- An insufficient power supply. Symptoms, include Windows 'bonging' regularly, hotspot rebooting, and no display or an incorrect display.

- The display is not configured correctly in Pi-Star. No display even while rebooting the hotspot.

- Incorrect Modem setting in Pi-Star. No transmission to the radio. No display.

- Hotspot being well off frequency. Seeing activity but no audio from the radio, no display of calling stations, hotspot not responding when you transmit.

- Being on the wrong frequency or offset. I was most confused when trying to transmit on my other hotspot's frequency.

- Network disabled in Pi-Star. Look for red warning bars on the pi-Star dashboard.

Loading Pi-Star

Almost all hotspots use the Pi-Star modem software. Its function is to manage the routing of data traffic received by the hotspot board to the YSF network. It also routes traffic from the YSF network to the hotspot board which transmits it to your handheld or mobile radio. There are at least two Pi-Star variants or 'forks' of Pi-Star. The EA7EE version supports static reflectors on YCS reflectors. The W0CHP version has a nicer desktop, better suited for wide the widescreen monitors we use these days. I believe that it also supports static reflectors on YCS reflectors, but I was unable to get that to work. These options are covered briefly on pages 119 and 45. I tried out both versions but decided to reinstall the standard version of Pi-Star.

Pi-Star is extremely capable, and if you get into the 'Expert Level' functions very complicated. We will start with a basic setup and proceed from there with caution. I only have a very superficial understanding of Pi-Star, "just enough to get me into trouble." This quote from the website sums it up nicely.

"The design concept is simple. Provide the complex services and configuration for Digital Voice on Amateur Radio in a way that makes it easily accessible to anyone just starting out but makes it configurable enough to be interesting for those of us who can't help but tinker." (Pi-Star UK).

Pi-Star runs on a Raspberry Pi or similar single-board computer. The dashboard interface is a web page that can be accessed using your favourite internet web browser. You make changes on the web page, and they are saved back to the hotspot. You can access the Pi-Star dashboard from your PC, tablet, iPad, or even a phone. Normally for a home hotspot, your controlling device has to be connected to your local LAN at home. You can configure Pi-Star for public access over the internet and you sometimes see this with publicly accessible repeaters and high-power hotspots.

Pi-Star includes the Raspberry Pi OS Linux distribution from the Raspberry Pi Foundation. It is a variation of Debian Linux that has been optimised to run on Raspberry Pi ARM boards. Check out the 'What is Pi-Star' page on the Pi-Star.uk website.

THE PI-STAR WEBSITE

The Pi-Star website is at https://www.pistar.uk/. Note that if you have a Pi-Star hotspot plugged in and you type 'Pi-Star' into the URL box, the computer may treat the entry as a search. You are likely to end up on your Pi-Star dashboard instead of the Pi-Star website. Pi-Star is based on the DStarRepeater and ircDDBGateway software designed by Jonathan Naylor G4KLX, which has been extended to support the full G4KLX MMDVM suite, including the cross-mode gateways added by José (Andy) Uribe, CA6JAU.

The website has a wealth of useful digital voice information as well as downloads for the latest revision of Pi-Star for various hardware platforms. The YSF section includes lists of the YSF and FCS reflectors.

HOTSPOT SD CARD

Unless your hotspot came with a fully programmed SD card, you will have to buy one. You can fit Pi-Star onto a 2 Gb micro SD HC card, but you probably will not be able to find one that small. I ended up with a 32 Gb card and 27 Gb of free space. The Raspberry Pi is not tremendously fast, so there is no need to buy a super-fast SD card. The card will probably already be formatted, but if not, you can format it with the FAT32 option. The image will overwrite the formatting with a Linux format anyway. SD cards are available from most electronics and computer shops, or the usual online sources. I have had good results using SanDisk cards, but any micro SD HC card should be fine. Kingston and Samsung disks are also recommended. I don't like Adata products. I had a bad experience with one of their SSDs.

TIP: If the card has already been formatted for Linux, Windows will flip out and try to open each partition. It will also ask you to format each partition. Just keep clicking the close button on all the popup windows until it quits asking. This is a real pain, and it will happen any time you put a Linux SD card into a Windows machine. If you happen to have a Linux machine you can bypass this problem by flashing the SD card on that.

Figure 40: A micro SD HC card and free adapter

SD card reader & writer

If you have a notebook PC, it may have a built-in SD card reader. If it does, it will probably be for the full-size SD cards, but you can buy a micro SD HC card that comes bundled with an adapter. If like me, you use a desktop PC without an SD card reader, you can buy a USB SD card reader. They write as well. They only cost $10 or £9 or thereabouts. You can use one that takes full-size SD cards and buy a micro SD card that is packaged with an adapter, or you can buy one that takes the micro SD card. I have one of each. One of them came free with a unit I bought from China. USB card readers are available from electronics and computer shops, or the usual online sources.

Downloading the Pi-Star image

Unless your hotspot came supplied with a pre-programmed SD card for the Raspberry Pi, the Pi-Star software has to be downloaded from the Pi-Star website. This is a big 612 Mb download.

It will require a good internet connection and it can take a long time, depending on your broadband speeds. The download took about 4 minutes at my place.

Open the Pi-Star.uk website in your favourite web browser and select **Downloads › Download Pi-Star** from the menu bar on the left. Or go straight to https://www.pistar.uk/downloads/. There are download options for several single-board computers. NanoPi, Nano Pi Air, Odroid XU4, and the Orange Pi Zero. There may be several releases for the Raspberry Pi listed. The RPi versions are OK for the Pi Zero and the Pi 3B. Download the newest release by clicking the orange text. The current release is **Pi-Star_RPi_V4.1.5_30-Oct-2021.zip.**

When the zip has finished downloading, unzip the files to a directory. Perhaps a Pi-Star directory or a Temp directory. It does not matter so long as you can find the files again. There are two files, a .img file which is the Pi-Star image and a .md5sum which is a checksum file used to detect whether the image file is corrupted.

Flashing the SD card

The website has a good set of instruction guides for flashing the image to an SD card, but I am going to step you through the process anyway. There is also a video at https://www.youtube.com/watch?v=B5G4gYDdJeQ.

There are a few programs that can write an image file to an SD card. I have tried 'Win32 disk imager' and 'Balena Etcher.' They both work well. I usually use Balena Etcher for Linux SD disks. It just seems to be easier to use and I am less likely to make a mistake. Balena Etcher warns you against selecting a large drive such as your PC hard drive. Download one or the other and install it on your PC.

"Send in the clones." If you have a friend with a working hotspot you might like to make a clone of their image file rather than download the latest version from Pi-Star.uk. You can use either program to make a clone of a working SD card onto a new SD card. Note that this is not the same as making a copy of the files. It copies the file structure and formatting as well. Technically it should be possible to format the card on a Linux machine and then copy the files, but I won't guarantee that won't end up causing you hours of tinkering.

Insert the SD card into your PC, either directly or in the USB to SD dongle. As discussed above you may have to use a micro SD to SD adapter as well. These come free with many micro SD cards.

Close any annoying error popup windows from Windows. Do not follow the advice to format the partitions.

Make 100% sure that you know the drive letter of the SD card. You do not want to write a Linux distribution onto a USB drive you happen to have plugged into your PC.

Balena Etcher method

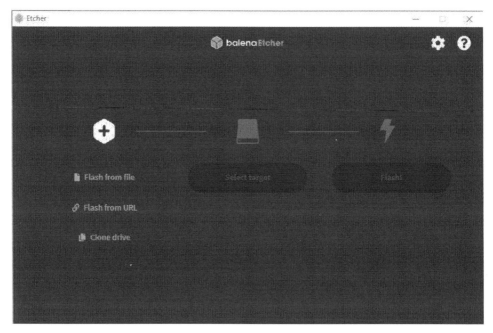

Figure 41: Balena Etcher

Click **Flash from file** and navigate to your .img file.

Click **Select Target**. It should find your SD card automatically but check the drive letter to make sure. Note the 'large drive' and 'source drive' warnings.

Select target	4 found			
Name	Size	Location		
⚠ WDC WD1003FZEX-00MK2A0	1 TB	G:\	Large drive	Source drive
⚠ Samsung PSSD T7 SCSI Disk Device	1 TB	E:\	Large drive	
☑ Mass Storage Device USB Device	32 GB	F:\		

Figure 42: Balena Etcher - choose the SD card

Click **Flash**. Wait until the write and validation cycle is complete. Close the Windows error windows, do not format the disk.

Remove the SD disk and the USB adapter from the USB port. Place the SD card into the SD card slot on the Raspberry Pi board.

Win32 Disk Imager method

Click the blue disk folder icon to the right of the top 'Image File' text field and navigate to the image file.

Use the 'Device' dropdown to select the SD card. Check the drive letter to make sure that it is the SD card and not some other drive.

Ignore all the other settings and click the **Write** button. The writing process will be shown on the progress indicator, then the validation process. It can be quite slow.

Wait until the write and verify is complete. Close or cancel the Windows errors. Do not format the disk. That would overwrite the information you have just placed onto the disk. Remove the SD disk and the USB adapter from the USB port.

Select the SD card. Note the warnings on the large drives. Check that the 'Size' matches the SD card and note the drive letter.

Click **Flash** and wait for the disk write and validation to complete.

Cancel any Windows errors and do not format the new disk volumes.

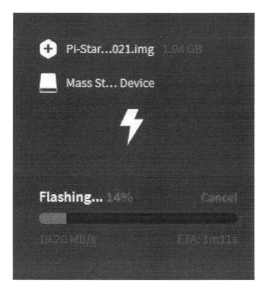

Figure 43: Win32 Disk Imager

LAN CONNECTION

Duplex hotspots running on Raspberry Pi 3 or 3B boards can be connected to your LAN with a direct Ethernet connection to your main router, or a hub. I have a four-port internet switch in the shack for connecting my HF transceivers and my PC. However, most people configure the hotspot for WiFi access, and I ended up using a WiFi connection. My duplex hotspot shows the IP address in the bottom right of the idle screen when there has been no talk group activity for a while.

Simplex hotspots always use WiFi because the Raspberry Pi Zero does not have an Ethernet connector.

WIFI CONNECTION

The SD card has been flashed, but the Raspberry Pi will not have access to your WiFi. Before you proceed you need to know the name of your WiFi network, which is referred to as the SSID (service set identifier) and the WiFi password which is called the PSK (pre-shared key).

TIP: This is the same information that you would use if you were adding a phone or computer to your WiFi network. If you don't know the WiFi name and password, it may be on the back or underside of your WIFI internet router. Or in the booklet that came with it.

If you are using a Raspberry Pi Zero W or Zero 2W, you need to complete the steps in configuration option 1 or option 2.

If you are using a Raspberry Pi 3 or 3B, the easiest method is to use configuration option 3. Which involves using the wired Ethernet connection to configure the WiFi connection. But you can use option 1 or option 2 if you like. If you plan to always use a wired Ethernet connection with your Raspberry Pi 3 or 3B, you don't need to set up a WiFi connection at all.

Option 1: Adding the WiFi settings using the Auto AP method

For this method, you need a computer, phone, or tablet with a WiFi connection. It avoids having to download a 'wpa_supplicant.conf' file and copy it into the boot directory of your new SD card. It is also useful if you take your hotspot somewhere and want to connect it to a different WiFi network.

Pi-Star can create its own WiFi network. When you power up the hotspot it will take a while to boot up. A Pi Zero could take up to a minute. If the hotspot is unable to connect to a WiFi access point within two minutes after it finishes booting up, it will create a WiFi access point (AP).

To configure WiFi access so that it can connect to your home network, you disconnect your PC or phone etc. from your usual WiFi network and connect to the Raspberry Pi access point instead. Then you configure the Raspberry Pi for your home network and reboot the hotspot. Lastly, you reconnect your PC or phone to your home network.

1. Boot up the hotspot and wait for two or three minutes until the AP has been created.

2. On a Windows PC, click the WiFi icon on the right side of the toolbar. On a phone, the WiFi settings are in the settings menu. Look for a new WiFi network called **Pi-Star-Setup and** connect to that. I did not have to enter a password, but if you do, it will be, **raspberry**.

3. You should be taken directly to the Pi-Star dashboard. If that does not happen, open your web browser and enter pi-star.local into the URL area. After a few seconds, the Configuration page should appear. Or click on **Configuration** top right.

TIP: Your web browser may want to treat pi-star.local as a search enquiry. If that happens enter http://pi-star.local or http://pi-star instead.

Thanks to W1MSG for his video https://www.youtube.com/watch?v=Z5svLP8nEyw

All done? Then skip ahead to the WiFi configuration instructions on page 108.

Option 2: Adding the WiFi settings using a manual setup

Pop back to the Pi-Star.uk website and select **Pi-Star Tools > WiFi Builder**. Enter your country code using the WiFi **Country Code** dropdown list. In the **SSID** box enter the name of your WiFi Network. In the **PSK** box enter your WiFi password, the same as if you were connecting a new phone or laptop to your network at home. Then click **Submit**.

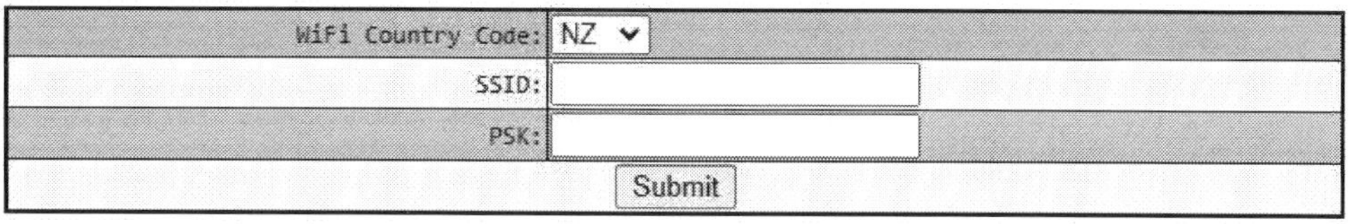

WiFi Country Code:	NZ ⌄
SSID:	
PSK:	
	Submit

Figure 44: Pi-Star.uk WiFi builder

*TIP: If you require a config that will connect to any available open network, leave the SSID and PSK lines empty, the generated config will allow your Pi to connect to any available **open** network. Then click on **Submit**.*

The website will download a small file called wpa_supplicant.conf to your PC.

Put the SD card in its dongle and plug it into your PC. The SD Card boot drive partition should open in Windows Explorer when you plug the SD card, into its USB dongle, back into the PC. Close or cancel any Windows error popups and do not accept the instructions to format the drive partitions. There will be a second drive as well, but you can ignore that. Copy or move the wpa_supplicant.conf file you downloaded into the SD card boot drive partition.

Done! Take the SD card out and insert it into the SD card slot on the end of the Raspberry Pi. The card will only go in one way, copper terminations end in first, with the card facing "up" towards the printed circuit board. It will slide in easily. There is no latch.

1. When the card has been inserted into the Raspberry Pi and the Pi-Star system boots up, it will add the config file for the WiFi and then reboot. There may be no indication on the hotspot that this has happened, and it may take a couple of minutes to sort itself out.

2. Open the web browser on your PC and enter pi-star.local into the URL area. After ten seconds, the Configuration page should appear. Or click on **Configuration** top right.

TIP: Your web browser may want to treat pi-star.local as a search enquiry. If that happens enter http://pi-star.local or http://pi-star instead.

Skip ahead to the Pi-Star Configuration chapter.

Option 3: Adding the WiFi settings wired Ethernet method

If you have a Pi3 or 3B, you can temporarily connect an Ethernet cable to your hotspot and set up the WiFi connection. If you have a situation where your only Ethernet port is remote from the shack, on your WiFi router, fiber, or ADSL modem, that is not a problem. You can configure the hotspot with the PC separated provided they are both on the same LAN. After the WiFi has been configured you can unplug the Ethernet cable and move the hotspot back into the shack.

Open the web browser on your PC and enter pi-star.local into the URL area. After a few seconds, the Configuration page should appear. Or click on **Configuration** top right. If you cannot connect to the hotspot, you will have to find the IP address of the hotspot and enter that as a URL. You can find the hotspot listed as 'Pi-Star' by accessing your main router, or by using the NetAnalyzer (iPhone) or FING app.

TIP: Your web browser may want to treat pi-star.local as a search enquiry. If that happens enter http://pi-star.local or http://pi-star instead.

Continue to the WiFi configuration instructions.

WIFI CONFIGURATION INSTRUCTIONS

If you used option 1 or option 3, you are not finished yet. If you used option 2 the hotspot should already be configured for WiFi, you can skip this section and proceed to the Pi-Star Configuration chapter. You should have seen the Pi-Star 'No Mode Defined' splash screen and after 10 seconds the Configuration page should have opened. If not, click Configuration at the top left of the Pi-Star screen. I will cover the general Pi-Star configuration in the next chapter. This section is just for completing the WiFi setup.

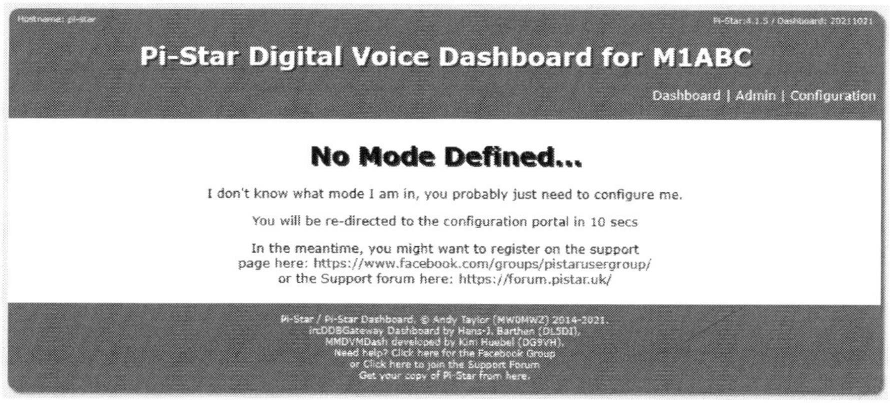

Figure 45: Initial Pi-Star screen

Scroll down the Config screen until you reach the 'Wireless Configuration' section. It should look like this. Note that it says that the WiFi interface is down. If it is green and says the WiFi interface is up, the WiFi is already configured and working.

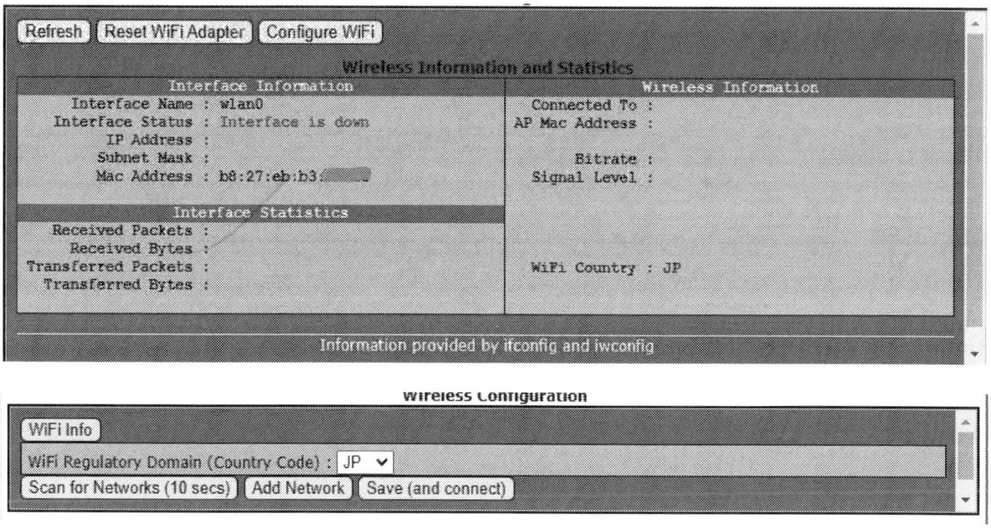

Figure 46: Pi-star WiFi Status and Config.

Just below the status box, there should be a 'Wireless Configuration' box, like this one. Select your country using the **WiFi Regulatory Domain (Country Code)** dropdown list. I selected NZ. Then press the **Scan for networks (10 seconds)** button. It should provide you with a list of WiFi sources that the hotspot can see. Your home LAN should be near the top because it should have the strongest signal. Select your WiFi LAN. This should populate the Wireless Config. With your Country Code and SSID (WiFi network name). Enter your WiFi password into the PSK box and press **Save (and connect)**.

Alternatively, you can set it up manually by pressing **Add Network**.

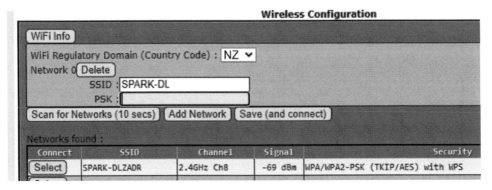

Figure 47: Choosing the WiFi network and entering your WiFi password

Above the 'Wireless Communication' box, there is a button marked **Apply Changes**. Click that. After about 30 seconds the Config screen will go back to the splash screen and then after another 30 seconds or so, it will return to the config screen. You should be rewarded with the green **Interface is up** message.

Take note of the IP Address in case you need it later.

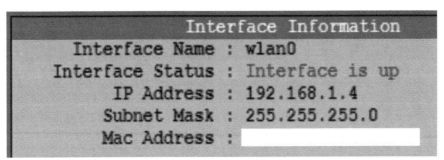

Figure 48: The WiFi is working :-)

IMPORTANT NOTE

Now you need to restart the hotspot to transfer from the AP mode (option 1) or the Ethernet mode (option 3) to WiFi. If you used option 1 now is the time to reconnect your phone/PC/tablet to your normal network. If you used the Ethernet cable, unplug it now. At the top of the page, there is **Power > Reboot > OK**. Wait 90 seconds for the reboot to finish, and the dashboard should reset. Or just power down the hotspot and restart it. The hotspot will jump to a different IP address when it restarts. It should be the one you recorded. That means that your Pi-Star web page may not work anymore. Retry the pi-star.local or http://pi-star.local page several times and it should eventually restart. If that does not work, try entering the IP address as a URL.

Pi-Star Configuration

OK, the hotspot is running, and you have configured the WiFi access, which is "the tricky part." Now Pi-Star has to be configured.

Pi-Star runs on the Raspberry Pi, but you configure it using a webpage interface on your PC, tablet, or phone. This is normally only accessible from within your home network. You can choose to make it public, but that would only be for a public repeater or high-powered hotspot.

If you are not already there, with the hotspot booted up and running and Pi-Star on your web browser, plug in the hotspot, start your PC web browser, and type

http://pi-star/ or pi-star or

http://pi-star.local/ or pi-star.local or

http://pi-star/admin/configure.php or

the IP address, usually something like 192.168.1.4

At least one of these options should bring up the Pi-Star dashboard.

PI-STAR ERROR MESSAGE

You may get an error message like this one. It just means that the modem selected in Pi-Star does not match your hotspot hardware, which is not surprising since you have not set that yet. Just click OK and carry on.

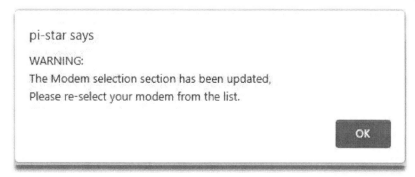

Figure 49: Pi-Star modem error message

PI-STAR BASIC CONFIGURATION

If you are on the dashboard page, click Configure to get started. If you are asked for a login name, it is pi-star and the password is raspberry.

WOW! This looks complicated. Luckily you only need to do a few things.

We are going to start with a basic connection and work up from there.

The top section is 'Gateway Hardware Information.' It contains the hostname, software kernel (revision), the platform (model of Raspberry Pi), CPU loading, and the CPU temperature. Green is good. Orange is OK. If it gets into the red, you need a heatsink or a fan on your Raspberry Pi CPU.

The next section is 'Control Software.' It should already have **MMDVMHost** selected because we are not configuring a D-Star repeater. If you have a simplex hotspot (one antenna) select **Simplex Node**. If you have a duplex hotspot (two antennas) select **Duplex Repeater**. This is **an important setting**.

The third section, 'MMDVM Host Configuration,' is where we set up different digital voice modes and transcoding between voice modes. We are only interested in YSF at present, so click the switch icon to turn **YSF Mode** on. The switch should change to an orange colour. We will leave the **RF Hangtime** and **Net Hangtime** set to the default 20 seconds. You can change them later if you want to. The last line in this section selects the display type if your hotspot has a display.

MMDVDM Display Type

If your hotspot has no display, set the MMDVDM Display type to **None**.

For an OLED (small display) select OLED type 3 for the very small 0.96' screen or **OLED type 6** for the more common 1.3" screen. Either option works on my hotspot.

If you have a Nextion screen, select **Nextion**.

There are a couple of other options, including one for TFT displays.

Port

For OLED displays, select **/dev/tty/AMA0** because the display is being driven directly by the modem board.

For Nextion displays you usually select **modem** unless the display is connected to a TTL to USB adapter plugged into the Rpi. In that case select **/dev/tty/USB0**.

Nextion Layout

This selects from four display layout options. Note that the background image is stored on the display itself. This dropdown only changes the information that is sent from Pi-Star to the display. It does not change the basic layout.

 The Nextion layouts can be edited on your PC and loaded directly into the Nextion display. You may have seen a power adapter in the Nextion box. See the information at. https://on7lds.net/42/nextion-displays.

Choosing the Nextion screen resulted in my display starting to show the IP address.

Click the **Apply Changes** button to load your changes to the hotspot.

The 'General Configuration' section includes some very **important settings**.

Identity

Hostname: **pi-star** I don't see any good reason to change this.

Node Callsign: This is usually your **callsign**. If it was a public repeater, it would be the repeater callsign.

CCS7 / DMR ID: If you have one, enter your (or the repeater's) **DMR ID number**. You don't need an ID number for YSF.

Radio Frequencies

Radio Frequency: If you selected 'Simplex Node' there will only be one frequency box. If you entered 'Duplex Repeater' and remembered to Apply Changes, there will be a receive frequency and a transmit frequency.

You can use any frequencies that are in the band(s) supported by both the hotspot and your handheld radio. Almost all hotspots operate in the 70cm amateur band. Although the hotspot transmits very low power and is unlikely the interfere with others, you must remember that your handheld may be transmitting a much higher power. Check your local 70cm band plan. There will be band segments for repeaters and digital simplex. For my simplex hotspot, I chose one of the designated digital simplex frequencies. I believe that it is a good idea to use the standard 5 MHz (NZ) repeater offset and band segment for a duplex hotspot or repeater on the 70cm band. I selected a repeater pair that is not in use for any other repeaters in New Zealand. Just choose a repeater frequency pair that is not being used in your region. If somebody complains it is a trivial matter to change to a different frequency, although it will mean changing the channel on your radio.

Setting		
Hostname:	pi-star	Do not add suffixes
Node Callsign:	ZL3DW	
CCS7/DMR ID:	1234567	
Radio Frequency RX:	438.125.000	MHz
Radio Frequency TX:	433.125.000	MHz

Figure 50: Pi-Star ID and frequencies

Remember that the duplex hotspot frequencies are the reverse of the frequencies that you program into your radio. The radio receives the hotspot transmit (TX) frequency and it transmits on the hotspot's receive (RX) frequency. I know that you already know this, but it is easy to get it wrong 😊.

Location

Add your **latitude and longitude**, using the degrees and decimal 172.1234 notation rather than degrees minutes and seconds. If you don't know your latitude and longitude, find your street on Google maps click on the map, then right-click, and your location will be displayed. Latitude is a positive number in the northern hemisphere and a negative number in the southern hemisphere. The longitude is measured in degrees East (+ve) or degrees West (-ve) of the zero line.

Enter your closest town or your locality and your Maidenhead grid. For example, Christchurch, RE66hm. Enter your **Country**.

URL you can enter your personal website or select **Auto** to use your QRZ.com listing. Most folks do that. Note that selecting Auto won't change the URL text box until you go through the **Apply Changes** process.

Latitude:	-43.497	degrees (positive value for North, negative for South)	
Longitude:	172.605	degrees (positive value for East, negative for West)	
Town:	Christchurch, RE66hm		
Country:	New Zealand		
URL:	https://www.qrz.com/db/ZL3DW	⦿ Auto ◯ Manual	
Radio/Modem Type:	MMDVM_HS_Hat_Dual Hat (VR2VYE) for Pi (GPIO) ⌄		
Node Type:	⦿ Private ◯ Public		
APRS Host Enable:	◯		
APRS Host:	aunz.aprs2.net ⌄		
System Time Zone:	Pacific/Auckland ⌄		
Dashboard Language:	english_uk ⌄		

Figure 51: Pi-Star location information

Radio modem type

The radio modem type must suit the hotspot design. You might have to experiment to find the correct one if the manufacturer didn't supply the modem type information. It will be an RPi one, not a Nano or USB stick. Mine is a 'dual hat' (duplex) BI7JT hotspot from China. The firmware was written by VR2VYE. The STM32-DVM / MMDVM_HS – Raspberry Pi Hat (GPIO) modem seems to work with most generic Chinese simplex hotspots. If you bought a ZUMspot try one of those options, if it is a DV-Mega try those options.

Node type

Set the node type to Private. It would only be set to public if your repeater or hotspot was also available publicly. If you do set it to 'public,' change the default password.

APRS Host

If you turn APRS Host Enable on, set the closest APRS host, or rotate.aprs2.net which guarantees a connection to a Tier2 APRS server.

It lets the hotspot announce its position on APRS.fi. I leave it turned off since my hotpot is not travelling anywhere.

System time zone and language

Set these to your time zone and preferred language option.

Click **Apply Changes**.

Yaesu System Fusion Configuration

The Yaesu System Fusion Configuration section sets the YSF network settings. It is much easier than setting up for DMR.

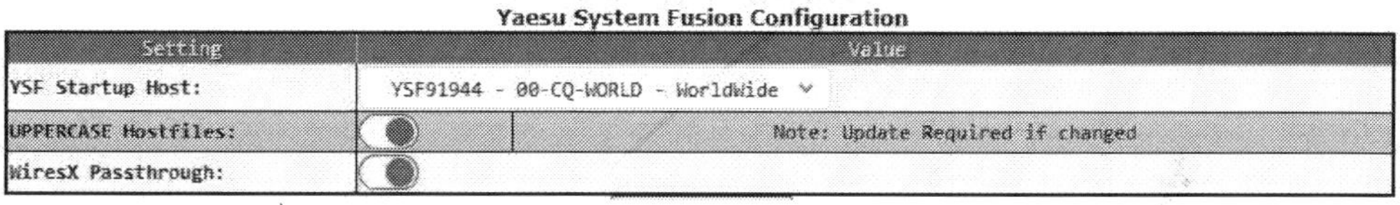

Yaesu System Fusion Configuration

Setting	Value
YSF Startup Host:	YSF91944 - 00-CQ-WORLD - WorldWide ∨
UPPERCASE Hostfiles:	◉ Note: Update Required if changed
WiresX Passthrough:	◉

Figure 52: Pi-Star YSF configuration

YSF Startup Host: can be set to start the hotspot linked to any YSF or FCS reflector. YFS41562 CQ-UK and YSF32592 America Link are the busiest by far.

Uppercase Host Files: enables a lowercase to uppercase conversion in the software. I believe that this is only necessary if you are using an FT-70D. But I leave it turned ON because you have to do a software update, not just the usual Apply Changes reboot if you change the setting. If you have an FT-70D make sure it is **ON**, if not it does not matter.

WiresX pass-through: must be enabled so that you can select reflectors from the radio. Leave it turned **ON**.

The transcoding modes are covered on page 88.

Mobile GPS

If you have a GPS receiver attached to the hotspot, (not the one in the radio), you can enable it, set the device port on the Raspberry Pi (Linux) and set its communication speed. I don't know what GPS data standard is required. I have not experimented with this option.

Firewall configuration

You can leave the first three options set to **Private**. Auto AP should be left **On**. It enables the AP mode if the hotspot can't find a WiFi connection. The uPNP setting should be left **On**. It enables 'universal plug-and-play router configuration, so you don't have to do manual port forwarding.

If you need access outside of your local LAN, over the internet, you can change any or all of the top three settings to **Public**. To quote Andy Taylor in the Pi-Star Users Support Group: "These settings tell the uPNP daemon to request port forwards from your router."

- Dashboard Access: requests TCP/80

- ircDDBRemote Access: requests UDP/10022

- SSH Access: requests TCP/22

SSH (Smart Shell) access will work over your local LAN when it is set to **Private**. Public access would only be used to let someone access your Raspberry Pi remotely, over the internet.

Auto AP SSID

This setting, under the WiFi Configuration, changes the default so that a password and login name is required to access the hotspot in AP mode. If you need the AP mode, you are probably in enough trouble without adding this extra layer of complexity. I suggest you leave it blank.

Remote access password

Use this if you have enabled **public** access over the internet. The public will, if they know the IP address or URL, be able to see the Pi-Star dashboard. This password stops them from re-configuring the hotspot. Leave this blank unless you are making the dashboard public.

PI-STAR BACKUP

When you have your settings right make a backup of the Pi-Star configuration. It will be saved to your Windows downloads area as a Zip file. Click **Configuration > Backup/Restore > Big down arrow**.

You can restore a previous arrangement by selecting the file with **Choose File** and clicking the **Big up arrow**. This could be useful if you make configuration changes that didn't work well, or if you want to save multiple arrangements. For example, you might have a setup that has two DMR networks and another that is configured for the YSF network. You can also use the zip file to load a Pi-Star configuration saved from another hotspot.

UPDATING PI-STAR SOFTWARE AND MMDVM FIRMWARE

WARNING: There is no need to update the Pi-Star software if you just downloaded it from the Pi-Star.uk website. It will already be the latest version. There will probably be no need to update the MMDVM hotspot firmware either unless it is a very old modem. Don't update either unless you have a good reason to do it.

Pi-Star software update

If you leave your hotspot plugged in and connected to the internet the dashboard will be updated nightly between 3:00 and 5:00 am. It will not update the version number for minor host file updates. It will update the version number if a program update is applied. You can also update Pi-Star from the Pi-Star dashboard. The version you are running is displayed at the top right of the dashboard page. You probably never noticed it. My dashboard is 'PI-STAR: 4.1.6 / Dashboard: 20221114. This means that the Pi-Star version is 4.1.6 and the Dashboard version was released on the 14th of November 2022.

To update the software, click **Configuration > Update**. A load of green text will show the Linux update happening on the Raspberry Pi. It takes a while and stops for up to two minutes at 'Updating PiStar-Firewall' during the process. "Be patient grasshopper," When the last line says **Finished**, re-boot the hotspot. The dashboard page will show the new version number if there was an update available.

SSH access to Raspberry Pi

You can access the Raspberry Pi that is running your hotspot with SSH (Secure Shell). There is an SSH shell built into Pi-Star. **Configuration > Expert > Tools SSH Access**.

SSH access using PuTTY

Most people use PuTTY for SSH access, but any SSH client will do.

I can't think of any reason to access the Raspberry Pi, but you can do a software update that way if you want to.

Enter the Pi-Star's **IP address** or **pi-star.local.** into the host name box. **Save** it see the note below. Then click **Open**.

I found that using the host name crashed PuTTY but the page would come up after a delay of a minute or more. It is probably something to do with my PC setup.

Using the IP address worked perfectly.

Figure 53: PuTTY SSH shell program setup

It is a good idea to enter the hostname or IP address and then save it by entering a name in 'Saved Sessions' and clicking the **Save** button. If you don't do that, this rather ominous warning message will pop up when you click Open. Just ignore it and click **Yes**.

Figure 54: PuTTY error message

SSH Login

In PuTTY you will be presented with a PuTTY terminal window. If you are using the built-in SSH you will not see the splash screen. It jumps straight to the login.

login as: **pi-star**
pi-star@pi-star.local.'s password: **raspberry**

Figure 55: SSH access to the Raspberry Pi

This screen indicates that you have access to the Raspberry Pi. You can use the normal Linux command lines, ls -a to list a directory, df -h to check the SD card space, etc.

Note the instructions on the boot screen that say 'Pi-Star's disk is read-only by default. You can change it to read-write access, at your own risk, using **rpi-rw**. You do not need to do this to run the update or upgrade commands. All pi-Star tools start with a **pistar-** prefix. (Note there is no hyphen between pi and star).

TIP: You can also get access to the Raspberry Pi operating system by plugging in a USB keyboard and an HDMI monitor and rebooting the hotspot.

Updating the Pi-Star software from the command line

There is no need to update the Pi-Star software from the command line because you can do it from the Pi-Star dashboard. But you can, using **sudo pistar-update**. I believe that it runs the same command as doing it from the dashboard, but it also updates the Raspberry Pi OS and other software on the Raspberry Pi. It can take a long time to complete. The update will stop for up to a minute on some lines. Just be patient, go and make a coffee, and let it run until you get the Linux prompt back.

You can also run **sudo pistar-upgrade** which will look for an upgraded version of Pi-Star. Well, that was interesting! When I ran the upgrade command, it said I was already running the latest version (as expected) but it now the dashboard reports that I am running version 4.1.6.

Updating the MMDVM firmware from the command line

It is unlikely that you will have to update your MMDVM modem firmware. I don't recommend doing it unless you have a compelling reason. "If it ain't broke, don't fix it." The MMDVM firmware revision is indicated in the Radio Info box on your Pi-Star dashboard. It is updated via SSH access using a command like **sudo pistar-mmdvmhshatflash hs_hat** or if you own a ZumSpot **sudo pistar-zumspotflash rpi**. You **MUST** know the correct firmware for your hotspot. It should be on the manufacturer's website. **If in any doubt don't do it.**

PI-STAR ALTERNATIVES

There are at least two alternative versions or 'forks' of Pi-Star. One is the EA7EE version that allows you to use static reflectors on the YCS network, another is the W0CHP version that replaces the traditional Pi-Star dashboard. Its 'Live Caller' screen is like the display I get on my hotspot's Nextion display, so this version would be a good option for a hotspot that only has a small display or no display at all. I tried out both versions but decided to reinstall the standard version of Pi-Star. There is also a phone app called Pi-Star mobile v1.40.

EA7EE version

The EA7EE Pi-Star software created by Manuel Sanchez Raya supports the YCS 'DG-ID' mode. You can select static modules on YCS reflectors. See the website at http://pistar.c4fm.es/. The download is at https://drive.google.com/file/d/1jU6Bia-DDkdH4bePjGDn_Iph6qMdOrts/view. There is a bit more about this on page 45.

The static reflectors work on a first in first served basis. If a call is heard on a module, it shuts out the other static modules until 20 seconds after the conversation has finished. You will transmit on the same module as you are hearing.

If you transmit when the hotspot is idle with no incoming calls you will transmit on the module you are linked to. It might be one of the static modules or a different one on the same YCS reflector.

If you have static modules set up in the EA7EE software, they will be active any time you link to an FCS reflector that is running on the YCS system. For example, FCS530 is also YCS530. The problem is that the modules on YCS530 which you may have set to static are not necessarily the same as the same module numbers on a different YCS reflector. So, you might end up listening to a completely different group. For example, on YCS235 America Link is on module 81 (FCS23581). But on YCS530, America Link is on module 37 (FCS53037).

TIP: EA7EE Pi-Star was a violet colour which didn't appeal to me. It was easy to set it to a nicer blue colour using the improved **CSS Tool** *in the Expert section.*

The EA7EE software supports EUROPELINK (YSF00007) and WORLDLINK (YSF00008) connections by passing through the DG-ID to the reflector. You can change 'rooms' on the reflector by setting the DG-ID on your transceiver to the room number. Or you can leave the DG-ID at TX 0, RX 0 and change rooms at http://worldlink.pa7lim.nl/ for Worldlink, or at http://europelink.pa7lim.nl/ for Europelink. These two reflectors were created by David PA7LIM. He also wrote Peanut and BlueDV.

W0CHP version

'W0CHP PiStar Dash' is available at https://w0chp.net/w0chp-pistar-dash/. There are heaps of screenshots and a video showing how easy the installation is. The software is easy to install and easily reversible back to the old Pi-Star, but please note there is **no support** other than the web page and the video. The software looks like it should support DG-ID linking and static DG-ID modules for YCS reflectors, but I was unable to get this to work. No doubt due to my extreme stupidity. Please read and understand the notes on the 'Musings' and 'Home' tabs before installing this software. I elected to uninstall it and return to using standard Pi-Star from pistar.uk.

Pi-Star Mobile v1.40

Pi-Star Mobile v1.40 is a patch for Pi-Star that lets you operate a 'mobile' version on your phone. You can also operate standard Pi-Star on your phone, but some of the text is rather small. The application is not freeware, but if you purchase the current version, you will get a free copy of version 2 when it is released. The current version works with iPhones and presumably Android phones. But it is only compatible with Pi-Star version 3, not the current Pi-Star version 4. I have not tried out this software. Because I didn't want to add a patch to my Pi-Star installation. There is a website at https://www.amateurradio.digital/pistar.php
and a video at https://www.youtube.com/watch?v=hmbPg-daCdQ&t=2s.

Troubleshooting

FTM-400XD, WIRES-X BUTTON NOT WORKING

The 'Wires-X' DX or GM/X button is not doing anything. It does not cause the radio to transmit, and the radio will not connect to the Wires-X, repeater, or hotspot. The most likely cause is that you are attempting to connect via the B-band on your radio. Wires-X only works on the A Band. On the FT5D, check the left side of the display for an orange square with an A in it. On the FTM-400XD check that the YSF repeater or hotspot frequency is on the top display

Also check, the radio is on the correct channel frequency and offset. That it is in the DN mode, and the DG-ID is the same as the node station or repeater. Almost always TX 0 and RX 0. The radio will not connect if any of those settings are incorrect.

RADIO KEYPAD BUTTONS NOT WORKING

The keypad may be locked. The FTM-400XD, FT5D, and FT-70D have a lock function on the **power** button. A short press will lock the keyboard, and another will unlock it. If your radio is different, see your radio manual for details on how to unlock it.

FTM-400XD NOT LINKING TO A HOTSPOT

If I turn on my FTM-400XD and set the hotspot channel, it should go into the connected mode when I long press the D/X button. For some reason, this often fails to work. You have to wait for the Wires-X symbol to disappear before transmitting. But after that, when I press the PTT, the transmission does not show up on the hotspot. Transmitting for a few seconds a couple more times usually clears the problem and I get a received indication on the hotspot. After that, I can connect. I don't have this problem with my other two YSF radios, or with my DMR and Icom radios, so I am pretty sure the problem lies with the radio.

FTM-400XD CAN'T CHANGE THE OPERATING MODE

You cannot change the operating mode FM, AM, DN etc. on the lower 'Band B' display. The radio will switch to AM automatically if you enter an Air Band frequency. Digital voice transmission including Wires-X is only possible on the upper 'Band A' display. You can listen to FM and FM, or digital and FM, but not digital and digital.

CAN'T HEAR THE REPEATER

You have your local repeater channel programmed, and you can see transmissions on the RSSI (receive signal strength indicator), but you cannot hear anything. If the repeater is a 'multi-mode repeater.' It could be transmitting a D-Star, P25, or DMR signal. You will only be able to hear the transmission if it is a C4FM or FM signal.

DIFFICULTY INSTALLING YAESU SOFTWARE

Join the club! I found the software installation to be extremely frustrating. Here are a few tips.

1. Re-install the USB driver. On the data cables, it is a 'Prolific' driver. The standard practice is to **not** have the cable plugged into the PC while you install the driver. However, the 'Prolific' driver installation in the Wires-X software installation process, specifically states that you should plug the cable into the radio and the PC before installing the driver. I had to install the USB drivers multiple times and reboot my PC when I installed the software for the ADMS programming software. If installing without the radio plugged in does not work, try again with the radio connected.

   a. Go into Windows Device Manager and look at the COM ports. For a data cable, you should see a COM port labelled 'Prolific USB-to-Serial Comm Port.' For a USB cable, you should see a port starting with YAESU, such as YAESU FT-70D Communication Device. The driver for the HRI-200 is labelled HRI-200 Communications Port. Take note of the allocated COM port number.

   b. If you do not see a COM port with a likely name, the driver has not been installed correctly. Check out Windows Device Manager, and COM ports. See which device disappears when you unplug the USB cable and reappears when you plug it back in. Try using that COM port number, but you probably will have little success with the software until the driver is correct. Try uninstalling the driver in Device Manager and reinstalling it.

2. Follow the installation instructions exactly. The process for using the FT-70 ADMS-10 software is downright medieval. See page 72.

3. If you get a 'Time out Error' the radio will go into an error mode. You will have to remove the power to get it to start again. Hold down AMS and plug in the power to get back to the ADMS mode. Check the COM port setting in the ADMS software. Select **Communications > COM Port Setting > Device Manager**. If the driver is installed correctly, the correct COM port will be labelled 'YAESU *FT-70D* Communication Device (COM_).' Click on that to select it and then press **Determine**.

CAN NOT CHANGE REFLECTORS/ROOMS

The usual reason is that the current reflector or the target reflector or Wires-X room is busy. You cannot unlink or link if someone is talking on the reflector. Wait for a gap in the conversation and unlink. Then link to the new reflector or room. Sometimes you can't link to a reflector without unlinking from the current one first.

CAN NOT CHANGE REFLECTORS - FT5D OR FTM-400XD

I had this situation happen when I set my hotspot to an XLX reflector. I selected 'YSF02401 – XLX299 – XLX Reflector.' When I put the FT5D into connected mode using the **GM/X** key, the display came up with XLX299 Reflector at the top where the hotspot is usually listed and Module R at the bottom. I found it very difficult to connect to another reflector from this state. However, on the FT5D, you can change to a different module on the reflector using

Search & Direct > ALL. On the FTM-400XD try the **white arrow > ALL.**

I could disconnect the reflector by holding down the **Band** button, as usual. But there was no indication on the radio that it had unlinked. I was not able to successfully use the search function to connect to a different reflector. Turning the **Dial** to link to a previous reflector worked, but again there was no indication of the change on the radio.

The best cure is to exit the connected mode, change the hotspot to a different reflector, and then set the radio back to the connected mode, using GM/X or DX.

OUTPUT POWER IS LOW

Some manufacturers are prone to exaggerating the RF output power. More specifically the radio probably will transmit at the stated power in FM mode if the battery is at full charge, but the output power may be much lower if the battery is down a bit. **Don't worry about it!** Remember that transmitting at half power is only a 3 dB reduction in signal strength at the repeater. So, unless you are right on the fringe, running slightly less transmitter power will not make any difference. The C4FM digital signal is wideband compared to unmodulated FM, so the RF power may look lower when you are transmitting a digital signal.

Use the lowest power setting if you are transmitting to a hotspot. You don't need full power to get across the room. For repeaters, use the lowest power setting that works reliably. Using the radio at full power makes it run hot and dramatically increases battery consumption.

HIGH BER INDICATED ON PI-STAR

A red high BER indication on the Pi-Star Local RF Activity and Gateway Activity screens, after you transmit to the hotspot, usually means that your hotspot is off frequency and an offset needs to be applied in Pi-Star. Not in your radio! See 'Hotspot off frequency,' below.

MY NEW HOTSPOT DOESN'T WORK

Don't panic! Most problems are easily solved. Sometimes just restarting the hotspot is enough. You are much more likely to have a problem with a simplex MMDVM hotspot than a duplex MMDVM hotspot. Other hotspots like OpenSpot or DV Mega, are less prone to problems and better supported by the manufacturer. The most common problems and symptoms are,

- An insufficient power supply. Symptoms, include Windows 'bonging' regularly, hotspot rebooting, and no display or incorrect display.

- The display is not configured correctly in Pi-Star. No hotspot display even while rebooting the hotspot.

- Incorrect Modem setting in Pi-Star. No transmission to the radio. No display.

- Network problems in Pi-Star.

- The hotspot is a long way off frequency. Seeing activity on the hotspot display but no audio from the radio, no display on the radio of calling stations, hotspot not responding when you transmit.

- The hotspot is a little off-frequency. Garbled incoming and/or outgoing speech. High BER or Loss % reported on the Pi-Star Gateway Activity screen.

Insufficient power supply

An insufficient power supply can be a problem, especially if you are using a Raspberry Pi 3 rather than a Raspberry Pi Zero W. I have had no issues powering my hotspot from a USB3 port on my computer, but the recommendation from the Raspberry Pi Foundation is to use a 2.5 amp USB power supply or a USB battery bank.

Display configuration

The display might not work if there is insufficient power, or it is not configured correctly in Pi-Star.

For an OLED (small display) select OLED type 3 for the very small 0.96' screen or OLED type 6 for the more common 1.3" screen. Either option works on my hotspot. The Port should be set to select /dev/tty/AMA0 because the display is being driven directly by the modem board.

If you have a Nextion screen, select Nextion. The port is usually set to modem unless the display is connected to a TTL to USB adapter plugged into the Rpi. In that case select /dev/tty/USB0.

Incorrect Modem setting in Pi-Star.

This is indicated by no transmission to the radio and/or no hotspot display. Check the advertisement for the modem you have purchased and see if there is any indication of the modem setting that should be set in **Configuration > General Configuration > Radio/Modem Type**. My duplex modem is set to **MMDVM_HS_Hat_Dual Hat (VR2VYE) for Pi (GPIO)**, as specified by the manufacturer.

Dual Hat means it is a duplex modem, VR2VYE is the person who wrote the firmware, it is for a Pi, and it is connected to the Pi using the GPIO header pins, not a USB interface.

I had trouble selecting the correct modem for my simplex hotspot, mostly because it was so far off frequency, it was not working anyway.

Make sure that you choose a Pi version, rather than a DV-Mega, Nano, or NPi version. Unless you do have one of those.

The **STM32-DVM / MMDVM_HS – Raspberry Pi Hat (GPIO)** modem seems to work with most generic Chinese simplex hotspots. Some of the other choices displayed calls on my hotspot, but there was no signal received by the radio.

Network problems in Pi-Star

Check the Pi-Star dashboard. The 'Modes Enabled' section should have **YSF** in green. The Network Status should have **YSF Net** in green. The last line shows the currently connected reflector, or **'Not Linked'** if no reflector is currently connected.

There are other indicators in the Service Status area, on the admin tab. Login with Username = **pi-star** and Password = **raspberry**. MMDVDM host should be green, indicating a Pi-Star hotspot is connected. YSFGateway should be green, indicating an internet connection through your home LAN to the YSF network, and PiStar-Watchdog should be green indicating that the Pi-Star software is regularly communicating with the hotspot. My hotspot has YSFParrot green as well.

If any of them are not green, go back through the Pi-Star setup.

Hotspot off frequency

This is the one that trips up most people, probably because it is the hardest to fix. It is very common for a new hotspot to be off-frequency. The frequency error should always be corrected in Pi-Star and never with the radio or programming software offsets. There is no point in deliberately making your radio transmit off frequency to compensate for the frequency error in the hotspot. How do you know it is the hotspot, not the radio? It is far more likely to be a cheap hotspot than an expensive radio. The only way to check for sure would be to compare it with another radio.

TIP: Some hotspots are checked by the vendor and shipped with an offset number on a slip of paper in the box or stuck to the bottom of the modem. I was not lucky in that respect.

When I powered up my new simplex hotspot it would not respond to my handheld radio when I transmitted, and although I could see callers on the hotspot display and a LED and RSSI indication on the radio, I could not hear them.

It wasn't until I watched a setup video on YouTube, I remembered the hotspot frequency offsets. I am fortunate to own a spectrum analyser and a good frequency counter, so I was able to quickly find out that my hotspot was a whopping 4.4 kHz off frequency. The online video was talking about a frequency offset of only 472 Hz.

First check

Before you attempt to change the hotspot offset, make very sure that the radio frequencies match the hotspot frequencies. Then check the Pi-Star Configuration tab.

If you are using a simplex hotspot, The Controller Mode in the Control Software should be set to **Simplex Node**. The Controller Software should be set to **MMDVM Host**. Both frequencies in the General Configuration should be the same. The radio channel should be set for simplex and should match the hotspot frequency.

If you are using a duplex hotspot, The Controller Mode in the Control Software should be set to **Duplex Repeater**. The Controller Software should be set to **MMDVM Host**. The frequencies in the General Configuration should be different.

Radio Frequency RX: is the hotspot's receive frequency. It should be set to the radio's transmit frequency. I use the normal repeater offset for the band. In New Zealand that is 5 MHz for the 70cm band or 600 kHz for the 2m band.

Radio Frequency TX: is the hotspot's transmit frequency. It should be set to the radio's receive frequency. This is the frequency programmed into the memory slot and displayed on the radio while it is receiving. The radio channel should be set for duplex and should match the hotspot frequencies.

If you still have a problem, you can set the frequency offset as described below.

Setting the frequency offset

The hotspot modem's radio uses a common oscillator for the transmitter and the receiver, so you can be pretty sure that any offset applied to correct the hotspot's transmit frequency will also be required to correct its receive frequency. I always set both offsets the same, and it always works.

There are two ways you can do this. Either adjust the receiver offset for the best BER (bit error rate) when you transmit to the hotspot. Or adjust the transmit offset while the hotspot is transmitting, by measuring the frequency it is radiating with a spectrum analyser, frequency counter, or SDR receiver. If the hotspot is seeing the transmission from your radio and the dashboard is showing 'Local RF Activity' when you transmit, use option 1. If the hotspot is not seeing the transmission from your radio and the dashboard is not showing 'Local RF Activity' when you transmit but is showing 'Gateway Activity,' when other stations use the connected reflector or repeater, use option 2. Option 2 is faster and more accurate, but it requires test equipment. Option 1 is a perfectly acceptable method. If you can't do option 1 or 2, try option 3.

Option 1: adjust the offset for the best BER

This method only works if the hotspot is seeing your transmission and it is showing up on the 'Local RF Activity.' If it is not and you do not have a frequency counter, spectrum analyser or SDR receiver capable of receiving the 70cm band you can try transmitting at a range of offsets until you can get into the hotspot.

Open Pi-Star and look at the 'Dashboard' page. Specifically, the BER (bit error rate) indication in the 'Local RF Activity' section.

1. Set your radio to your hotspot frequency. Unlink from any reflectors.

2. Key up your radio with the PTT and hold in on transmit for four or five seconds. You should see your callsign pop up in the Local RF Activity area and at the top of the Gateway Activity region. When you release the PTT, you should see an indication on the BER meter (hopefully <2%) and the SRC box should be green with 'RF' in it. If it is green the frequency offset is good. Job done. If it is red, the frequency offset needs to be adjusted.

    To adjust the offset in Pi-Star, select **Configuration > Expert > MMDVM Host**. Scroll down to the Modem section and find **RXoffset** and **TXoffset**. (Not RXDCoffset or TXDCoffset).

3. Set both **RXoffset** and **TXoffset** to 200 Hz. Click **Apply Changes**.

4. Transmit again after the hotspot resets and note if the BER got better or worse.

    a. If it got worse, the frequency may be high. Set both **RXoffset and TXoffset** to -200 Hz and try again.

    b. If it got better, increase the offset and try again. Keep changing both offsets, using smaller and smaller changes, until the BER is less than 2%.

Option 2: adjust the hotspot using its transmitter frequency

Using a frequency counter or spectrum analyser is more accurate than using the BER method, but it does require you to have the test equipment.

Remember to use a 20 dB coaxial RF attenuator in the cable if you are connecting your hotspot to a frequency counter or spectrum analyser.

1. Select a busy reflector like America Link, so that there is plenty of activity on the channel.

2. Take note of the hotspot's transmitter frequency (or simplex frequency). It is in the Radio Info box on the Pi-Star dashboard.

3. In the Pi-Star software select **Configuration > Expert > MMDVM Host**. Scroll down to the Modem section and find **RXoffset** and **TXoffset** (not RXDCoffset or TXDCoffset).

4. Monitor the hotspot and observe the frequency when it transmits. Subtract the wanted frequency as indicated on the Pi-Star dashboard from the observed frequency.

5. Adjust both the **RXoffset** and **TXoffset** by the amount of offset needed in Hertz. Click **Apply Changes**. After the hotspot reboots, observe the frequency when the hotspot transmits. Make any minor adjustments. You should be able to get to within about 100 Hz. The adjustment process is not fine enough to get the frequency exactly right.

6. You should make a final check that the BER is good when you transmit to the hotspot, using the instructions in option 1: The BER indication should be less than 2% and green.

If you don't have a frequency counter or spectrum analyser, you can use an SDR receiver to do the test. I do not recommend a direct connection. Attach an antenna to the SDR. It should be sensitive enough to see the signal from the hotspot. The frequency accuracy of your SDR should be good enough to determine if the modem is transmitting a long way off frequency.

Option 3: The hard way

If you don't have test equipment and the modem is not seeing your transmission, i.e. the 'Local RF activity' is not going red while you are transmitting, you can use the 'trial and error' method. Setup as per option 1. Set a 500 Hz offset on both the **RXoffset** and **TXoffset** click **Apply Changes** and see if you can hear traffic on the hotspot or if it displays your callsign when you transmit to it. No joy? Try a -500 Hz offset. Keep stepping up (or down) in 500 Hertz steps until you hopefully have some success. I think it is unlikely the hotspot will be more than 5000 Hz high or 5000 Hz low. That is 10 steps up from zero and 10 steps down from zero. Once Pi-Star indicates that it can see your transmission and the 'Local RF activity' goes red while you are transmitting. You can try the steps in option 1. If you get lost, go back to zero offsets and try again using 200 Hz steps.

FTM-400XD STANDBY BEEP

Is it just me or is the beep every time someone stops talking extremely irritating? I can hear when someone stops talking. I don't need a beep to tell me! Turn it off with, Hold DISP >TX/RX > Digital > Standby Beep > OFF.

FTM-400XD BEEP

I don't need a beep to remind me that I just touched a screen icon or a button. Unfortunately turning it off means you don't hear a chime when you are connecting to a room or a reflector, but you can't have everything! Turn the beeps off with, Hold DISP > CONFIG > BEEP > OFF.

Glossary

59	Standard (default) signal report for amateur radio voice conversations. A report of '59' means excellent readability and strength.
73	Morse code abbreviation 'best wishes, see you later.' It is used when you have finished transmitting at the end of the conversation.
.dll	Dynamic Link Library. A reusable software block that can be called from other programs.
2m, 70cm	Two metre (144 MHz) and 70cm (430 MHz) amateur radio bands
A/D	Analog to digital
ADC	Analog to digital converter or analog to digital conversion
AF	Audio frequency - nominally 20 to 20,000 Hz.
Algorithm	A process, or set of rules, to be followed in calculations or other problem-solving operations, especially by a computer. In DSP it is a mathematical formula, code block, or process that acts on the data signal stream to perform a particular function, for example, a noise filter.
AMBE+2	Advanced Multi-Band Excitation version 2. The AMBE+2 Vocoder uses a propriety chip made by DVSI (Digital Voice Systems Incorporated) to convert speech into a coded digital signal or the received digital signal back to speech. Can transmit intelligible speech with data rates as low as 2 kbs. Used for D-Star, DMR, YSF, NXDN, and 'phase 2' P25.
APCO	Association of Public-Safety Communications Officials
APRS	Automatic packet reporting system – used to send and display location information from a GPS receiver. APRS beacons are displayed on the APRS.fi website.
BER	Bit error rate – a quality measurement for any digital transmission system. It measures the number of bits that were received incorrectly compared to the overall bit rate.
Bit	Binary value 0 or 1.
bps	Bits per second (data speed)
BW	Bandwidth. The range between two frequencies. For example, an audio passband from 200 Hz to 2800 Hz has a 2.6 kHz bandwidth.
C4FM	Continuous 4-state Frequency Modulation (used for P25 Phase 1 and Yaesu System Fusion)
Carrier	Usually refers to the transmission of an unmodulated RF signal. It is called a carrier because the modulation process modifies the un-modulated RF signal to carry the modulation information. A carrier

	signal can be amplitude, frequency, and/or phase modulated. Then it is referred to as a 'modulated carrier.' The output of an oscillator signal is not a carrier unless it is transmitted.
CODEC	Coder/decoder - a device or software used for encoding and decoding a digital data stream.
CPU	Central processing unit. The ARM (advanced RISC machine) processor in the Raspberry Pi, or the microprocessor in your PC. [RISC is reduced instruction set computing, an acronym inside an acronym.]
Cross-connect	A link between different technologies or networks. Such as a D-Star reflector linked to a System Fusion 'Room' or a DMR 'Talk Group'
CTCSS	Continuous Tone Coded Squelch System, used for access control to most analogue FM repeaters and FM handheld or mobile radios.
D/A	Digital to analog.
DAC	Digital to analog converter or digital to analog conversion
Dashboard (Pi-Star)	The Pi-Star dashboard is an HTML website hosted on the Pi-Star hotspot and accessed via any web browser on the same WiFi or local network. It displays traffic being passed through the hotspot and the hotspot's configuration settings.
Dashboard (repeater, reflector, or Wires-X room)	A repeater or reflector dashboard is an HTML website that displays the status of a repeater or reflector, including what frequencies it is on, its location, who owns it, who is using it, and what reflectors it is linked to. Most repeater dashboards are available to the amateur radio community, or selected people, over the internet.
data	A stream of binary digital bits carrying information
dB, dBm, dBc, dBV	The Decibel (dB) is a way of representing numbers using a logarithmic scale. Decibels are used to describe a ratio, the difference between two levels or numbers. They are often referenced to a fixed value such as a Volt (dBV), a milliwatt (dBm), or the carrier level (dBc). Decibels are also used to represent logarithmic units of gain or loss. An amplifier might have 3 dB of gain. An attenuator might have a loss of 10 dB.
DC	Direct Current. The battery or power supply for your radio, charger, or Hotspot will be a DC power supply.
DMR	DMR stands for Digital Mobile Radio. It is an ETSI (European Telecommunications Standards Institute) digital voice standard employed by a wide variety of manufacturers. Motorola and Hytera are the biggest commercial vendors of DMR radios and trunked radio repeater systems. Motorola calls their version of DMR, MotoTRBO.
DMR+	DMR+ is a worldwide DMR network. It was the first to interconnect ETSI standard Tier II repeaters. It is aligned with the DMR-MARC

	network so you can access the DMR-MARC Talk Groups. The DMR+ network specialises in interconnections with other technologies such as D-Star, AllStar, and C4M (P25 and YSF).
DMR-MARC	DMR-MARC is a network of DMR repeaters established by the Motorola Amateur Radio Club. The members of MARC were instrumental in getting DMR established for amateur radio. They set up the first amateur radio DMR networks and repeaters. There are around 500 DMR-MARC repeaters in 83 countries with over 144,000 registered users.
D-Star	Digital Smart Technologies for Amateur Radio. D-Star is the (mostly) Icom digital voice system. Unlike DMR it was developed specifically for amateur radio.
DVSI	The AMBE+2 Vocoder uses a propriety chip designed and made by DVSI (Digital Voice Systems Incorporated) to convert speech into a coded digital signal or the received digital signal back to speech.
Duplex	A radio or Hotspot that can receive and transmit at the same time. Usually on different frequencies. A standard repeater is a duplex system.
DX	Long-distance, or rare, or wanted by you, amateur radio station. The abbreviation comes from the Morse telegraphy code for 'distant exchange.'
Echo	A D-Star reflector, usually the E extension, that repeats back a test transmission that you make. See Parrot.
ESSID	Extended service set identifier – a 2-digit extension to your DMR ID number to identify a second or subsequent hotspot on the same network.
ETSI	The European Telecommunications Standards Institute.
FM	Frequency modulation. The "good ol'" analogue repeater system.
FSK	Frequency Shift Keying. YSF transmitters use 'continuous' 4-state frequency shift keying modulation. Each frequency shift carries two bits of the input data stream
FTDI	USB to 3.3V TTL level converter designed by Future Technology Devices International Ltd.
GM	Digital Group Monitor. Checks which stations in a nominated group are within communication range. Displays the range and direction to those stations.
GMSK	Gaussian Minimum Shift Keying - spectrum efficient frequency shift keying mode used for D-Star
GPS	Global Positioning System. A network of satellites used for navigation, geolocation, and very accurate time signals.

Hex	Hexadecimal – a base 16 number system used as a convenient way to represent binary numbers. For example, 1001 1000 in binary is equal to 98h or 152 in decimal.
Hotspot (YSF)	A YSF Hotspot is a small internet-connected box that can connect to YSF and FCS reflectors. You transmit from your C4FM handheld to the Hotspot, and it passes the data through to the internet. The information that is returned is transmitted by the Hotspot back to your radio. Most YSF hotspots are MMDVM (Pi-Star). The rest are mostly OpenSpot or DVMega hotspots.
Hotspot (WiFi)	Many cell phones can be configured to act as a WiFi hotspot, allowing WiFi devices to get access to the internet via your phone and mobile data plan. You could connect a WiFi-enabled YSF hotspot to a WiFi hotspot on your phone and connect your YSF radio to worldwide reflectors via your phone.
Hz	Hertz is a unit of frequency. 1 Hz = 1 cycle per second.
IMBE	Improved Multi-Band Excitation. Vocoder used for P25 phase 1 digital voice
JARL	Japan Amateur Radio League. The official Japanese amateur radio organisation. Developed D-Star in association with Icom.
kbits	Thousands of bits per second. 1 kbit – 1000 bps.
kHz	Kilohertz is a unit of frequency. 1 kHz = 1 thousand cycles per second.
LAN	Local Area Network. The Ethernet and WIFI-connected devices connected to an ADSL or fibre router at your house are a LAN.
LED	Light Emitting Diode
LoTW	Logbook of the World. An ARRL QSO logging database which is used worldwide.
Mbits	Millions of bits per second (data rate)
MHz	Megahertz – unit of frequency = 1 million cycles per second.
MIC	Microphone
MMDVM	Multi-mode digital voice modem - usually supports DMR, D-Star, YSF, P25, and NXDN.
Module	FCS reflectors and YCS servers have up to 100 modules that link to various rooms and reflectors. The Worldlink YSF00008 and Europelink YSF00007 PA7LIM reflectors have up to 99 modules (known as rooms).
MOTOTRBO	MOTOTRBO is a Motorola trademark used to describe their range of DMR products
Network (YSF)	The YSF network is a collection of interconnected repeaters and reflectors. It includes YSF and FCS reflectors, some of which are hosted on the YCS servers.
Onboard	A feature or data list that is contained within the radio.

Parrot	A colourful class of birds known for their ability to mimic speech and other sounds. In YSF it is a system that repeats back a test call made from your radio. Known as Echo on D-Star.
PC	Personal Computer. For the examples throughout this book, it means a computer running Windows 10.
PCB	Printed circuit board
PSK	Pre-shared key – a password or security code known to your router and a connected router. For example, your BM password.
PTT	Press to talk - the transmit button on a microphone – pressing the PTT makes the radio transmit.
QSO	Q code – an amateur radio conversation or "contact."
QSY	Q code – a request or decision to change to another frequency.
PDN	Portable digital node. A way of using a YSF radio as a data node for Wires-X access. It has Terminal or Access Point modes.
POTA	Parks on the Air
RF	Radio Frequency
RS232	A computer interface used for serial data communications.
RPi	Raspberry Pi single-board computer
RX	Abbreviation for receive or receiver
SBC	Single board computer, such as a Raspberry Pi.
Simplex	Simplex means to receive and transmit on the same frequency. In most cases, a radio operating on a simplex frequency cannot transmit and receive simultaneously. Simplex can be used if you wish to communicate directly with another YSF radio without using a repeater or the internet.
Simplex Repeater or Hotspot	A simplex repeater uses the same frequency for receiving and transmitting. It passes data (digital voice) from an internet connection to the hotspot or repeater transmitter so that you can receive it on your radio. When you transmit, the simplex repeater or hotspot passes the signal received from your radio to a talk group over the internet connection.
SOTA	Summits on the Air
Split	The practice of transmitting on a different frequency to the one that you are receiving on. Repeaters use a 'repeater split' or 'offset' between the repeater input frequency and the repeater output frequency.
Squelch	Squelch mutes the audio to the speaker when you are in FM mode and not receiving a wanted signal. When a signal, with the correct CTCSS tone or sufficient signal strength is received, the squelch 'opens' and you can hear the station.
SSID	Service set identifier – in Pi-Star it is your WiFi network name

Tail	Furry attachment at the back of a dog or cat. Also, the length of time an FM repeater stays transmitting after the input signal has been lost. It can also mean a short flexible length of coaxial cable at the antenna or shack end of your main feeder cable.
Talk Group	A Talk Group is the equivalent of a YSF 'reflector' or a Wires-X 'room.' It is a collection of linked repeaters that are configured so that users with a common interest or from a common location can talk to each other. For example, there are Worldwide, North American, State, and County Talk Groups. Also, Spanish Language and Old Timers groups.
TDMA	Time domain multiple access. A technique for interleaving the data from two or more voice or data channels onto a single data stream. DMR uses TDMA to combine two voice (and data) Time Slots onto a data stream, which is used to modulate the radio using 4FSK.
TX	Abbreviation for transmitting or transmitter.
UHF	Ultra-High Frequency (300 MHz - 3000 MHz).
USB	Universal serial bus – serial data communications between a computer and other devices. USB 2.0 is fast. USB 3.0 is very fast.
USB	Upper sideband of a single sideband transmission.
VFO	Variable Frequency Oscillator. This applies to radios that can be tuned in frequency steps rather than stepping through previously saved memory channels.
VHF	Very High Frequency (30 MHz -300 MHz)
Vocoder	A Vocoder is a category of codec that analyses and synthesizes the human voice signal for audio data compression, multiplexing, voice encryption or voice transformation. The vocoder was invented in 1938 by Homer Dudley at Bell Labs as a means of synthesizing human speech. [Wikipedia]
W	Watts – unit of power (electrical or RF).
Wires-X	Wide coverage Internet Repeater Enhancement System. It is a Yaesu-owned and operated internet-based, reflector system. Not to be confused with the YSF reflectors reached via a DV dongle, Peanut, hotspot, or non-Yaesu repeater.
YSF	Yaesu System Fusion. You should know this by now!

Table of drawings and images

Internet links

The internet is quite big. I can't mention the thousands of links to all of the possible YSF references. Here are a few that I found useful.

Active Wires-X rooms https://www.yaesu.com/jp/en/wires-x/id/active_room.php

Wires-X software website https://www.yaesu.com/jp/en/wires-x/index.php

Hotspot software and a lot of other reference information https://www.pistar.uk/

Reflector dashboard links and other information http://xreflector.net/

YSF Reflector ID numbers https://register.ysfreflector.de/

YSF Reflector ID numbers https://www.pistar.uk/ysf_reflectors.php

FCS reflectors https://www.pistar.uk/fcs_reflectors.php

YCS reflectors. Link to DVMatrix in the left panel at http://xreflector.net/

YCS reflector Wiki http://ycs-wiki.xreflector.net/doku.php

YCS235 dashboard http://ycs235.xreflector.net/ycs/#

YCS530 dashboard http://ycs530.xreflector.net/ycs/#

Pi-Star available from https://www.pistar.uk/

Pi-Star download https://www.pistar.uk/downloads/

EA7EE Pi-Star version http://pistar.c4fm.es/

EA7EE Pi-Star download
https://drive.google.com/file/d/1jU6Bia-DDkdH4bePjGDn_Iph6qMdOrts/view

PA7LIM Treehouse project https://www.pa7lim.nl/treehouse/

PA7LIM WOLRDLINK reflector dashboard http://worldlink.pa7lim.nl/

PA7LIM EUROPELINK reflector dashboard http://europelink.pa7lim.nl/

PA7LIM BlueDV download http://www.pa7lim.nl/bluedv-windows/

PA7LIM Peanut download http://www.pa7lim.nl/peanut/

List of Peanut rooms http://peanut.pa7lim.nl/rooms.html

APRS.fi website https://aprs.fi

NXDN talk groups https://www.pistar.uk/nxdn_reflectors.php

A website devoted to C4FM radios and more http://c4fm.xyz/radios-c4fm.php

Great videos

Many people have created excellent videos relating to YSF. Sometimes seeing someone perform a task is easier than reading about it. Here are a few that I liked. If you like a video on this list, please hit the 'Like' button and subscribe to the channel. It encourages the creators and helps pay for more great videos.

Build your own DMR/DStar/Fusion hotspot for CHEAP – YouTube
https://www.youtube.com/watch?v=LspgnvDPJvc

A video by Matt M6CEB demonstrates the setup for operating P25 from a YSF radio and your hotspot. https://www.youtube.com/watch?v=IYPrQ93xro4&t=1s

A video by 'radiosification', demonstrates the setup for operating P25 from a YSF radio and your hotspot. https://www.youtube.com/watch?v=IC9r4_LQLk8

RT Systems ADMS software https://www.youtube.com/watch?v=RRy-TO588sk

Flashing an image to an SD card https://www.youtube.com/watch?v=B5G4gYDdJeQ

Pi-Star WiFi setup W1MSG https://www.youtube.com/watch?v=Z5svLP8nEyw

Pi-Star mobile v1.40 https://www.youtube.com/watch?v=hmbPg-daCdQ&t=2s

M0FXB how to change FT5D from Europe B2 to UK C2 repeater format.
https://www.youtube.com/watch?v=EGAeUTZ5ktA

M0FXB Yaesu FT5D Hints and Tips. M0FXB has made dozens of great videos!
https://www.youtube.com/results?search_query=MoFXB+yaesu+videos

Yaesu UK Change the Yaesu FT5DE Region to UK
https://www.youtube.com/watch?v=CCBFj1XNmKs

David Hunter Yaesu Ft-991A Wires-X tutorial
https://www.youtube.com/watch?v=p7UyXw0Q69o

Episode 2 – Wires -X This is one in a series of great videos from Yaesu USA Official
https://www.youtube.com/watch?v=FPb2iHF3GdE

Episode 3 – HRI-200 setup, another great video from Yaesu USA Official
https://www.youtube.com/watch?v=WtNjsRXkrK0

Setting up a WIRES-X node - Part 1 – North West Fusion Group
https://www.youtube.com/watch?v=8G7ihF1sJa0

Setting up a WIRES-X node - Part 2 – North West Fusion Group
https://www.youtube.com/watch?v=7TAFPQHabwA

Index

The Author

Well, if you have managed to get this far you deserve a cup of coffee and a Gingernut biscuit. It is not easy to digest large chunks of technical information. It is probably better to dip into the book as a technical reference. Anyway, I hope you enjoyed it and that it has made learning about Yaesu System Fusion a little easier.

I live in Christchurch, New Zealand. I am married to Carol who is very understanding and tolerant of my obsession with amateur radio. She describes my efforts as "Andrew playing around with radios." We have two children and two cats. James graduated from Canterbury University with a degree in Commerce and is working for a large food wholesaler. Alex is a doctor working at Christchurch Hospital.

I am a keen amateur radio operator who enjoys radio contesting, chasing DX, digital modes, and satellite operating. But I am rubbish at sending and receiving Morse code. I write extensively about many aspects of the amateur radio hobby.

Thanks for reading my book!

73 de Andrew ZL3DW.